THE MECHANISM
OF
CREATIVE EVOLUTION

CHARLES DARWIN   GREGOR MENDEL

THE MECHANISM
OF
CREATIVE EVOLUTION

BY
C. C. HURST

Doctor of Science and Doctor of Philosophy of the University of Cambridge;
Sometime Fellow Commoner and Research Student of Trinity College,
Cambridge; Fellow of the Linnean Society of London;
Author of *Experiments in Genetics*

WITH
A FRONTISPIECE AND
197 FIGURES

CAMBRIDGE
AT THE UNIVERSTY PRESS
1933

CAMBRIDGE
UNIVERSITY PRESS

University Printing House, Cambridge CB2 8BS, United Kingdom

Cambridge University Press is part of the University of Cambridge.

It furthers the University's mission by disseminating knowledge in the pursuit of education, learning and research at the highest international levels of excellence.

www.cambridge.org
Information on this title: www.cambridge.org/9781107495234

© Cambridge University Press 1933

First edition 1932
Reprinted 1933
First published 1933
First paperback edition 2015

A catalogue record for this publication is available from the British Library

ISBN 978-1-107-49523-4 Paperback

CONTENTS

PREFACE

WE live in an age of scientific research when important discoveries are made every day and there is a danger lest, in the multiplicity of these discoveries, essential synthetic values may be lost in a maze of detail. After thirty years' experiments carried out in different parts of the world, with representative species of every type of living organism, the time seems to be ripe for a gathering together of the multitudinous facts which go to make up the genetical story of creative evolution. Future discoveries will no doubt enrich the story and fill in many details which are now missing, but the main mechanism of the genes and chromosomes which constitute the physical basis of creative evolution, has already been experimentally established, as shown in this book. Mendel's original discovery, developed by the experiments of Bateson and others and crowned by the brilliant work of Morgan and his colleagues, has led to consequences so far-reaching as to bring about one of the greatest revolutions that science has known, and has resulted in the elevation of biology to the rank of an exact science. The new science of genetics has confirmed and extended Darwin's law of natural selection and Mendel's laws of heredity, both experimentally and mathematically. One of these extensions is the genetical delimitation of species which has enabled us to place the study of creative evolution and its natural adjunct taxonomy, on the experimental basis of an exact science.

This century has also witnessed equally remarkable and revolutionary discoveries in the physical sciences and pure mathematics, which have given us entirely new views of the constitution of matter and of the universe around us. The discoveries of the chromosomes and genes and their inter-relations prove to be as important to the biologist as the discoveries of the atoms and electrons and their inter-relations are to the physicist. The gene is the unit of life and the genetical species is the unit of creative evolution. The far-reaching importance of these vital units to mankind can hardly be overestimated, since the genes are not only the basis of all structural and functional characters but, as recent work indicates, they are also the foundation of human thought and action.

It is evident that as a result of these discoveries, a critical period in human history has arrived. Man, if he chooses, can, here and now, take a hand in creative evolution by creating new species of living organisms and replacing natural selection by human selection. For a thousand million years natural selection, as one of the processes of creative evolution, has dominated life, and for ten million years it has dominated the human mind, obsessing Man with the idea of an overruling fate. Scientific research has brought freedom to Man, and the future trend of creative evolution, including Man's own destiny, depends entirely on his response to the new knowledge and on his intelligent application of these discoveries in the near and distant future.

My grateful thanks are due to genetical colleagues in various parts of the world, and to the publishers specified below, for their generosity in allowing me to reproduce illustrations. I am also much indebted to my wife who has helped me in my researches and has illustrated them in drawings which appear in this book.

C. C. H.

CAMBRIDGE
1932

Acknowledgments are due to the following for leave to reproduce illustrations: Messrs Edward Arnold & Co.; The Bobbs-Merrill Co.; California University Press; Cambridge University Press; Cassell & Co., Ltd; Chicago University Press; Columbia University Press; Herr Ferdinand Enke, Stuttgart; Herren Gebrüder Borntraeger, Leipzig; Harvard University Press; Messrs Headley Brothers, Ltd; Henry Holt & Co., Inc., U.S.A.; The J. B. Lippincott Co., U.S.A.; The McGraw-Hill Book Co., Inc.; Macmillan & Co., Ltd; The Macmillan Co., U.S.A.; N. V. Martinus Nijhoff's Boekhandel's-Gravenhage; Messrs Oliver & Boyd; Oxford University Press; Princeton University Press; Yale University Press.

INTRODUCTION

EVOLUTION

IN its original meaning evolution represented a rolling out or unfolding, as in the development of an opening leaf or flower bud of a plant or the emergence of an animal from the egg. In this sense it was almost synonymous with the modern word ontogeny which signifies the somatic development of an individual organism from a spore, egg or embryo to its full development, decline and death. In other words, evolution was individual. In the eighteenth century, when the preformation theory of Bonnet was in the ascendant, evolution was used strictly in this sense and the idea persisted up to the time of Lamarck in the early nineteenth century. It is evident that the Lamarckian theory of the inheritance of acquired characters was based largely on the old ideas of individual evolution and variation. Eventually in Darwin's time evolution came to mean any changes, or "transformations" as they were called, in living organisms, whether individual or racial. Since the time of Darwin, the use of the word evolution has been extended far beyond its early meanings and has been applied in all directions. It is necessary therefore to point out that the evolution dealt with in this book is limited to organic evolution or more precisely the evolution of living organisms. The qualification "living" is necessary in view of Whitehead's recent application of the term "organism" to non-living matter. In this limited sense, evolution simply represents the origin of new kinds of living organisms from old ones.

CREATIVE EVOLUTION

The phrase "*L'Évolution Créatrice*" originated with the French philosopher Henri Bergson in his charming literary account of evolution. Naturally, as a philosopher, Bergson deals with the subject from the point of view of human values rather than from a biological standpoint. With the eye of a philosopher he perceives that evolution is, in the main, progressive. With the eye of a naturalist, the biologist observes that the evolution of living organisms as a whole represents a progressive system of increasing structural, functional and genic complexity, from the minute sub-cellular *Bacterium*, with its pre-cellular responses, to the multi-cellular mammal and primate *Homo*, with his reflective and conceptual mind. While neither accepting nor rejecting Bergson's vitalistic and orthogenetic theory of "*L'Élan vital*" but regarding it as redundant, many biologists have recognised that Bergson's description of evolution as "creative" is a true and apt expression of the biological facts. Before definitely adopting Bergson's literary and philosophical phrase as a scientific term, however, it is necessary that it should be more precisely defined in a scientific sense. For genetical biology a verbal definition is inadequate and it is necessary that there should be a quantitative and qualitative measure of creative evolution which can at any time be submitted to the test of genetical experiment. Recent work shows that in the genetical species we have such a unit which, though not entirely free from objection, is probably the most useful one available in the present state of knowledge. Since the time of Darwin the "good" taxonomic species has been used as a rough and crude measure of evolution, but experience teaches that this is a concept too vague and subjective to be used as a measuring rod of precision in a scientific age.

The genetical species proves in general to be a considerably larger unit than the ordinary Linnean species, being in many cases equivalent to a generic section or a sub-genus. It is measurable in terms of chromosomes and genes and consequently can at any time be precisely determined and tested experimentally. In this respect it provides a convenient and satisfactory biological unit of creative evolution which may be simply defined as the origin of new genetical species from old ones.

MECHANISM

The word "mechanism" in the title of this book has been used with some hesitation and doubt, since the word is always open to misconstruction when applied to living organisms. It is, however, a useful word to express the fact of the regular and orderly sequence of processes found in all healthy living organisms. So long as it is clearly understood that no ulterior materialistic meaning is attached to the word, no difficulty should arise. Evolution is not a deterministic machine, and creative evolution, as we shall see, is far from being mechanistic—on the contrary it is peculiarly indeterminate throughout all its phases. At the same time the chromosomes which play a great part in creative evolution have a definite visible mechanism of their own in the different stages of mitosis and meiosis which cannot be mistaken, and experiments show that the genes associated with them are equally concerned in this mechanism. The genetical mechanism is of course a living one or at least it is concerned with living entities, but it is none the less a mechanism expressing the regularity and orderliness of the activities and processes involved, in the same way that we speak of physiological, psychological and other biological mechanisms.

HISTORICAL

The idea of organic evolution is more ancient than is generally supposed, but in the nineteenth century of our era, the genius of Charles Darwin raised it to a high peak in popular thought. The speculations of the early Greek philosophers display a prophetic insight of modern ideas of evolution.

In the sixth century B.C., Anaximander conceived the idea of a gradual evolution from chaos to order and the transformation of aquatic species to terrestrial species by adaptation.

In the fourth century B.C., Empedocles believed that plants were evolved before animals and that imperfect forms arose by random combinations and were gradually replaced by perfect forms through the extinction of those which could not support themselves and multiply.

In the third century B.C., Aristotle attributed all evolutionary changes to natural causes. He rejected the hypotheses of their fortuitous origin and the survival of the fittest, and favoured the idea of intelligent design in the formation of adaptive characters. He graded evolution in a linear series from plants, sponges, sea-anemones and animals up to man. He anticipated Harvey's doctrine of epigenesis in development and duly noted the facts of heredity, dominance and reversion. In the first century B.C., Lucretius the poet followed Empedocles and ignored

Aristotle, he believed that structure preceded function as a random variation, and inferred the existence of heredity bodies within the mother. Aristotle's views of evolution, however, held the field for over two thousand years and were revived by Francis Bacon in the sixteenth century of our era. Bacon also added new ideas of the mutability of species by the accumulation of variations and large mutations and believed that new species arose from old species by a process of degeneration.

It is now apparent that the ecclesiastical opposition to Darwin in the last century was largely due to the unconscious influence of Aristotle's teaching which had persisted throughout the Dark Ages and was handed on indirectly through the early theologians.

In the eighteenth century, ideas of evolution became more prominent. Buffon believed in the influence of the environment in modifying the structures of animals and plants and in the inheritance of these acquired characters. He also emphasised the struggle for existence among animals and plants.

Born in the same year, the great systematist Linnaeus apparently believed in the special creation of genera, but thought that species and varieties originated by hybridisation between different genera. In the same period Bonnet expounded the novel idea that the Creation described in Genesis was only a resurrection of animals previously existing.

It was about this time that Erasmus Darwin, the grandfather of Charles Darwin, anticipated Lamarck in his ideas of the internal origin of adaptive characters. He thought that animal species were transformed by their own exertions in response to pleasure and pain and that many of these propensities were transmitted to posterity. He recognised the beneficial struggle for existence which checked too rapid increase of life. He was the first to make clear that organic evolution had persisted for millions of years and that all life arose from one primordial protoplasmic mass.

The similar views of his contemporary Lamarck had a much wider influence at that time and have even survived to this day. Early in the nineteenth century Geoffrey St Hilaire strongly supported the idea that evolution was caused by the direct action of the environment inducing profound changes in the egg, thus anticipating Weismann's idea of germinal variations. St Hilaire also believed that new species might be formed in this way from sudden large variations in one generation, thus anticipating the mutations of de Vries.

The way was therefore well prepared for the work of Charles Darwin who in 1859 in his *Origin of Species* was the first to provide an adequate proof of the facts of evolution and whose theory of natural selection is now accepted by biologists as a primary law of nature. Darwin was the first evolutionist to employ the inductive method of research by first observing and collecting facts and from these constructing working hypotheses to fit them. No one man, before or since, has amassed so many facts and observations as Darwin did and the sheer weight of the evidence he produced was so convincing that it overwhelmed all opposition to his views of evolution. Darwin's chief contribution was his law of natural selection which may be briefly stated as follows:

(1) Individuals of a species vary at random in many directions.

(2) These variations are transmitted to offspring.

(3) Those individuals which vary in a more favourable direction, fitting them to live in changed conditions, survive and multiply in larger numbers.

(4) Consequently in each generation there is a slow but definite approach to complete adaptation to different conditions of life.

(5) In secular time, as conditions change considerably, new species evolve and become established which are adapted to the new conditions and in this way evolution becomes an orderly procession of species adapted to the varied and variable conditions of life such as we find in Nature.

During and since the time of Darwin there have been many attempts to replace his law of natural selection with alternative theories of evolution. These have taken on Protean forms but most of them have a common basis in postulating with Aristotle, Erasmus Darwin, Goethe and Lamarck, that variation is not indeterminate in many directions, as Darwin found, but determinate in definite directions according to the needs of the species. In other words, orthogenesis replaces natural selection in the evolution of adaptations and species. Examples of modern orthogenetic theories are found in Carl von Nägeli's inner principle of progressive development, Eimer's orthogenesis, Cope's bathmism and kinetogenesis, Driesch's entelechy or vitalism, Bergson's *Élan vital* or the urge of life to creative evolution, and most recent of all, Berg's nomogenesis which may be described as a form of internal predestination. It will be observed that all these orthogenetic theories from Aristotle and Lamarck to Bergson and Berg, involve a mystical factor or principle which, in so far as it is incapable of experimental identification and verification, is beyond the province of science and must therefore, so far as biology is concerned, be regarded as purely speculative and relegated to the realm of metaphysical philosophy where it may be more usefully discussed. Incidentally it may be pointed out that the scientific mechanism of the chromosomes and genes expounded in this book goes far to explain a large number of the facts commonly attributed to the principle of orthogenesis by systematists, palaeontologists and embryologists.

On the whole the definite opponents of Darwinism among biologists are a small minority. Naturally with the progress of knowledge many new facts have come to light since the publication of Darwin's *Origin of Species* and considerable modifications and adjustments of Darwin's early views have been necessary. These modifications at the time appeared to be in direct opposition to Darwin's views, but as time passed it became more and more evident that these modifications have served to strengthen Darwin's original position. Space will allow only a few of these to be mentioned. For instance, Weismann supported the idea of natural selection but showed conclusively in his theory of the germ plasm that acquired characters are not as a rule inherited, as was assumed by Lamarck and Darwin and their contemporaries. Johannsen experimentally demonstrated the difference between fluctuating varieties or modifications which are not inherited and germinal variations or mutations which are. It may be pointed out, however, that although fluctuating modifications and acquired characters do not appear to be specifically inherited as such, yet they are potentially inherited, since they usually represent different reactions of the same gene complex in different environments and tend to be repeated in the same environment.

Bateson, with his critical and sceptical mind which sifted everything, was one of the first to realise the discontinuous nature of species and the difficulties of Darwinism, yet in the end Bateson's work in creating the new science of genetics and developing Mendelian experiments in plants and animals, paved the way for the new genetical Darwinism which has already thrown so much light on the nature of species and evolution.

The genetical approach to Darwinism has been further strengthened by the mathematical work of Fisher and Haldane, which has placed the study of natural selection on a higher plane, and has provided a new tool and approach to the problem, from which much is expected during the next decade. Fisher's work on *The Genetical Theory of Natural Selection*, with its new views of the origin of dominance and the natural selection of genotypes, has already become a classic. J. B. S. Haldane's *Mathematical Theory of Natural Selection* shows that in evolution, neither mutation nor Lamarckian transformation can prevail against natural selection of even moderate intensity. In America, Sewall Wright has also made an extensive mathematical investigation of the problem of evolution and natural selection, and in the main his results agree with those of Fisher and Haldane although he attaches more importance to random survivals in medium-sized populations than either Fisher or Haldane.

Darwin and Wallace both made some notable contributions to the important question of the geographical distribution of species and the evidence it provides for natural selection. Since their time this problem has been dealt with on a large scale both by Willis and by Vavilov. Willis in his *Age and Area* believes that a new species often arises from an old one by a sudden process of mutation of many characters and that although it survives by natural selection, yet natural selection is not a cause of the new species appearing. He presents important statistical evidence that rare plant species of restricted habitat are of recent origin and estimates that in the case of flowering plants about two species per century arise suddenly as new mutations. On Willis's theory each new species forms a centre of distribution and its area of distribution increases with the age of the species. Genetical experiments with Willis's rare and abundant species of a single genus (*Coleus*), that grow side by side, would no doubt solve the problem of their origin and test his theory which is supported by a general survey (statistically treated by Yule) of plants and animals in widely scattered areas and islands.

Confirmation of Willis's theory of *Age and Area* is also found in the remarkable work of Vavilov who has investigated the origin and distribution of the varieties of many species on an enormous scale in different continents. Like Willis he finds that species originate in primary and secondary centres and gradually extend their areas. This applies to both wild and cultivated species and varieties. He finds that dominant genes are concentrated in the centres of distribution while recessive genes are more prevalent around the periphery of the area. Vavilov and his army of Russian geneticists have already tested a large number of genes in these varieties in a large number of species from different parts of the world and in a few years a harvest of results should follow which will throw considerable light on the evolution of species. Another aspect of the problem of distribution and origin is found in some interesting statistical studies of thirty-five species of night-flying moths which have been carried out by Fisher and Ford and which show that species that are abundant are more variable than those that are rare. Rare species tend to decline and abundant species to increase. The explanation seems to lie in the rarity of advantageous mutations, which have a greater chance of establishing themselves if the breeding population is large. A large number of less abundant species will generally be decreasing, while a smaller number of those more abundant will be increasing, the total number of species being maintained by fission of the more abundant rather than of the less abundant types. The extensive work of Turrill in the geographical distribution of species in conjunction with the genetical experiments of Marsden-Jones has proved to be of primary importance in increasing our general knowledge of the nature and genic constitution of genetical species.

Many attempts have been made to supplement Darwin's law of

natural selection and to clear up the outstanding difficulties in Darwin's views of heredity and variation. In the nineteenth century Mendel, Bateson, de Vries, Miss Saunders and a few others, foresaw that the only way to achieve this was by experimental investigations on strict scientific lines. From these small beginnings the new science of genetics arose in the first decade of the present century and in the course of thirty years it has completely revolutionised biological thought and has established research in evolution on the experimental basis of an exact science. The incorporation by Morgan, at an early stage, of the branch of cytology known as karyology, was an inspiration, since it has provided a new approach to the experimental study of evolution which has brought about the experimental demonstration of the living mechanism of heredity, variation and creative evolution. The concept of a genetical species experimentally based on combined studies in taxonomy, cytology and genetics has provided a unit of creative evolution and foreshadowed a new taxonomy based on genetical experiments.

The genetical creation of a new species of *Nicotiana* in 1925 by Clausen and Goodspeed and the similar creation of a new genus *Aegilotricum* in 1926 by Tschermak and Bleier, opened a new era in the experimental study of species and evolution.

The experimental creation by X-radiation of mutational and transmutational forms of *Drosophila* by Muller in 1927 proves to be another landmark in creative evolution, the ultimate issue of which cannot yet be estimated.

During the last few years much valuable cyto-genetical work has been done at the John Innes Horticultural Institution at Merton under the able leadership of Sir Daniel Hall, and systematic investigations of the species of *Rosaceae*, *Liliaceae* and other large families are being made.

The object of this book is to point out the salient features of these recent genetical discoveries and to show that the mechanism of the chromosomes and the genes therein is the mechanism of creative evolution.

CHAPTER I

MENDEL'S LAWS OF HEREDITY

IN 1857 an Austrian monk named Gregor Johann Mendel commenced a remarkable series of experiments in his cloister garden in the Abbey of Brünn. Interested in the problems of heredity and evolution, he began to experiment with the ordinary Garden Pea (*Pisum*). With a stroke of genius Mendel concentrated his experiments not only on one species of plant but on single characters in large numbers of individuals through several generations. The garden pea is a plant eminently suitable for this purpose, most of the races breeding true to type and many of the characters being distinct and discontinuous, that is to say, they do not grade into one another but are obviously different to the most casual observer. Moreover, as the plant is an annual, a generation can be raised in one year and large numbers can be raised in a small garden. Mendel selected seven pairs of the most distinct characters with which to work, and for illustration we will take one of these.

Most races of peas can be divided into two definite kinds, those with green and those with yellow seeds. Mendel first crossed a true-breeding yellow-seeded pea with pollen from a green-seeded pea and the result in the first generation (F_1) was all yellow-seeded peas (fig. 3). He then changed round and crossed a green pea with pollen from a yellow pea but the result was still the same, and in the first generation he obtained all yellow peas. He tried this experiment each way many times, always with the same result, pure yellow peas appeared every time in the first generation, and, although they had a green pea either for a father or a mother, no trace of the green parent could be found in the first crosses.

Mendel then sowed these cross-bred yellow peas and allowed them to self-fertilise. Each of the plants in the second generation (F_2) bore seeds of two colours, yellow or green, both colours often being found in the same pod (fig. 3). This proved that the cross-bred yellow peas contained the green factor although it was not patent, so Mendel called the yellow character *dominant* and the green character *recessive*. The next step was to sow the yellow and green peas of the second generation. Mendel sowed the green-coloured peas and found that they produced nothing but green peas in the third generation (F_3), and these being sown again gave

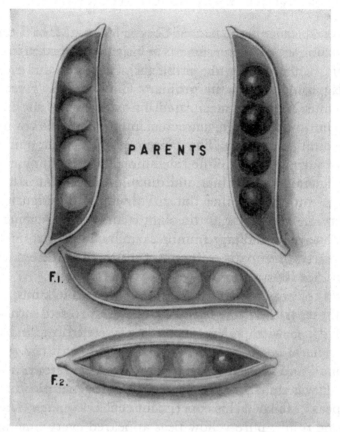

Fig. 3. The result of crossing Yellow (light-coloured) and Green (dark-coloured) Peas. In F_1 all yellow peas appear, in F_2 three yellows to one green. (After Morgan.)

nothing but green peas and so on for many succeeding generations $(F_4...F_n)$. In a word, the green peas were "fixed" and always bred true with no throwing back to the yellow colour, notwithstanding that both parents and two grandparents were the dominant yellow. The yellow peas of the second generation, however, behaved very differently. When

Mendel sowed these he found that they were of two distinct kinds. Some were also "fixed" and bred quite true without throwing any greens but others threw a mixture of yellows and greens, just as the yellow peas from the original cross had done. He worked on for many generations with these peas and always found that while the greens and part of the yellows bred absolutely true the remainder of the yellows would always produce three kinds—pure yellows, pure greens and the impure yellows which would again give the same mixture. Throughout the experiments

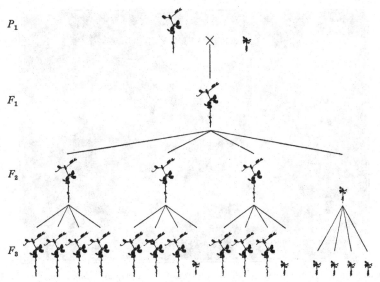

Fig. 4. The inheritance of tall and dwarf Peas. F_1 gives all tall (dominant), F_2 three talls to one dwarf. Of these in F_3 one tall breeds true, the other two throw talls and dwarfs in the ratio of 3 : 1, and the one dwarf breeds true. (After Darbishire.)

the pure yellow peas were quite indistinguishable from the impure yellows and their true nature could only be determined by their breeding behaviour. The purity of the extracted yellow dominants and the extracted green recessives was quite unexpected and forms the basis of what are now known as the Mendelian principles of heredity.

Another important and interesting fact that came to light during the experiments was that the three different types of peas always appeared in certain definite numerical proportions. On the average, out of every four peas there were one pure yellow, two impure yellows and one pure green, i.e. 25 per cent. pure yellow : 50 per cent. impure yellow : 25 per cent. pure green.

In Mendel's day it was known that at fertilisation an egg-cell of the female parent was joined by a pollen- or sperm-cell from the male parent, the union of these two forming the embryo from which the new individual arises. Mendel thus conceived the idea that these cells in some way bore the inherited characters, and that the yellow peas gave off germ-cells carrying the necessary factor to produce yellow, while the

Fig. 5. Inheritance in the Tobacco Plant (*Nicotiana*). (Left) *N. alata grandiflora* with large flowers; (right) *N. Forgetiana* with small flowers; between, four flowers from the F_2 generation, showing the normal 3 : 1 ratio in the segregation of the characters, large and small flowers. (After Morgan.)

green peas carried the factor to produce green. When these two types of peas were crossed together the germ-cell from the yellow pea carrying the yellow factor met the germ-cell carrying green from the green pea and the two together formed an individual whose cells carried both yellow and green factors, but since in this case yellow is dominant to green, only yellow is visible in the first cross. The subsequent appearance of green peas in later generations proved that although it did not appear in the first cross the green factor still retained its identity and individuality, ready to appear at any time when by chance it became once more free of the dominating influence of the yellow factor. Mendel conceived that when the first cross formed its germ-cells these factors segregated, one-half of the germ-cells carrying green and one-half yellow. In normal chance fertilisations the green factor should become free from the dominating influence of the yellow on the average once in four times. Since the first cross, which carries equal numbers of pollen- and egg-cells bearing the yellow and green factors, is self-fertilised, yellow should meet green and green meet yellow in 50 per cent. of cases giving 50 per cent. impure yellows, while yellow should meet yellow in 25 per cent. of cases giving 25 per cent. pure yellows and green should meet green in 25 per cent. of cases giving 25 per cent. pure greens. The visible result therefore would be, on the average, 75 per cent. yellow and 25 per cent. green, or a ratio of 3 : 1 in the second generation (F_2). Mendel's actual experiments gave 6022 yellow and 2001 green seeds, or a ratio of 3·01 : 1. On

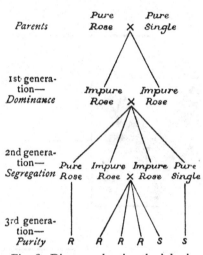

Fig. 6. Diagram showing the inheritance of Rose and Single Comb in Poultry.

repeating Mendel's original experiment the author obtained 1310 yellow and 445 green, or a ratio of 2·94 : 1. Other repeats by Correns, Tschermak, Bateson, Lock and Darbishire added to the above give a total of 134,707 yellow and 44,692 green, or a ratio of approximately 3·01 : 1.

The simple idea of the unity and purity of factors in the germ-cells, or gametic purity, is the most important of Mendel's discoveries and on it are founded all the complicated calculations of the probabilities of inheritance in different matings. This is known as the First Law of Mendel, the Law of Segregation of Germinal Units (now called "genes").

So far we have dealt with the inheritance of one pair of characters alone, now let us see what happens when more than one character pair is involved. Another difference between the seeds of garden peas is that

(a)

Fig. 7. Photographs of the progeny of a cross between recessive White Bantam cocks and dominant Black hens (a), and the reciprocal cross in which a Black cock is crossed with a White hen (b).
In both cases the F_1 generation shows only black birds, whether cocks or hens, black being dominant

(b)

to white in Bantams, and in the F_2 we get three black birds to one white in both cocks and hens, thus showing that whichever way the cross is made there is no difference in the behaviour of the dominant character Black and the recessive White. (After Jull and Quinn, *Journal of Heredity*.)

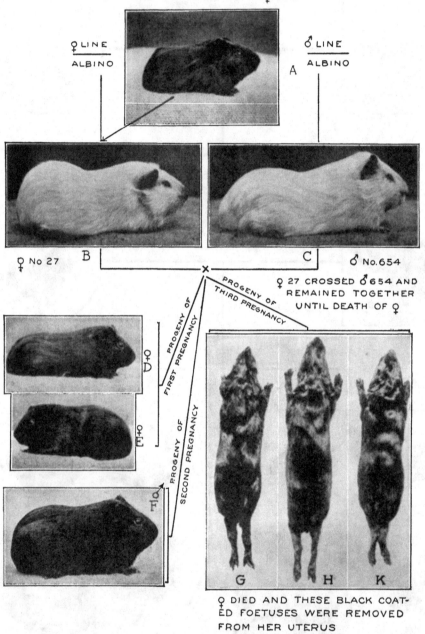

YOUNG ♀ PURE BLACK LINE AND SOURCE
OF TRANSPLANTED OVARIES IN ♀ No. 27

♀ LINE
ALBINO

♂ LINE
ALBINO

A

♀ No 27   B

C   ♂ No.654

X

PROGENY OF THIRD PREGNANCY

♀ 27 CROSSED ♂ 654 AND
REMAINED TOGETHER
UNTIL DEATH OF ♀

PROGENY OF FIRST PREGNANCY

♀ D

♀ E

♂ F

PROGENY OF SECOND PREGNANCY

G   H   K

♀ DIED AND THESE BLACK COAT-
ED FOETUSES WERE REMOVED
FROM HER UTERUS

Fig. 8. Demonstration of the ability of the germ-cells to remain pure even in cases where they
are nourished by unlike body-cells. Ovaries from a pure black guinea-pig were engrafted into
a pure breeding albino which, on being mated with a pure albino male, produced only black
offspring for three generations (until its death). (After Castle.)

in some races they are round and smooth while in others they are an-
gular and wrinkled. We will now consider the crossing of a round yellow
pea with a wrinkled green pea. In the first generation all the cross-bred
seeds are round yellow, whichever way the cross is made, because, as
Mendel found, the round and yellow characters are dominant to the
corresponding wrinkled and green characters, which are recessive. In

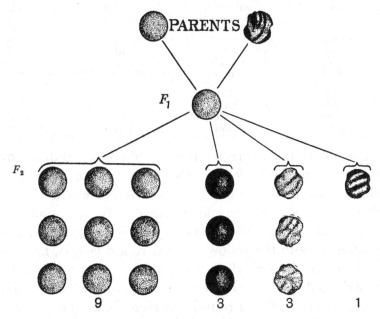

Fig. 9. The segregation of two distinct characters in Peas, round wrinkled and yellow green,
showing in F_2 on the average, 9 round yellow, 3 round green, 3 wrinkled yellow and 1 wrinkled
green. (Parents = round yellow × wrinkled green, F_1 = round yellow.) (After Morgan.)

the second generation these round yellow cross-breds self-fertilised
produce on the average a proportion of 3 round to 1 wrinkled and 3
yellow to 1 green. Taking the two characters in combination, there are,
in every 16 peas, on the average 9 round yellow, 3 round green, 3
wrinkled yellow and 1 wrinkled green or $(3\,R:1\,W) \times (3\,Y:1\,G)$
$= 9\,RY:3\,RG:3\,WY:1\,WG$. Fig. 9 and the following table will
explain this more clearly.

Here we see that the round yellow-seeded first cross has four kinds of
germ-cells, those carrying factors for round and yellow, those carrying

		Egg-cells of the Round Yellow Cross		
	RY	*WY*	*RG*	*WG*
RY	*RY* *RY* round yellow (pure)	*WY* *RY* round yellow	*RG* *RY* round yellow	*WG* *RY* round yellow
WY	*RY* *WY* round yellow	*WY* *WY* wrinkled yellow (pure)	*RG* *WY* round yellow	*WG* *WY* wrinkled yellow
RG	*RY* *RG* round yellow	*WY* *RG* round yellow	*RG* *RG* round green (pure)	*WG* *RG* round green
WG	*RY* *WG* round yellow	*WY* *WG* wrinkled yellow	*RG* *WG* round green	*WG* *WG* wrinkled green (pure)

(Left margin label: Pollen-cells of the Round Yellow Cross)

wrinkled and yellow, those carrying round and green and those carrying wrinkled and green. These four kinds occur in approximately equal numbers both in the pollen- and in the egg-cells. Since yellow and round are dominant to green and wrinkled respectively, the resulting individual shows the dominant characters only, and out of sixteen individuals only one is free of both the yellow and the round factors, i.e. a double recessive which will be wrinkled green. With regard to the others, of the three wrinkled yellow seeds one breeds true, which carries only wrinkled and yellow with no round or green. The other two split up again into wrinkled yellow and wrinkled green. Of the three round green peas one alone breeds true—that which carries only the round and green, the other two split up again into round green and wrinkled green. The nine round yellow peas are more complicated still. Only one—that carrying only round and yellow—breeds true. Of the others, two throw round yellows and wrinkled yellows, two throw round yellows and round greens, while the remaining four are like the original cross-bred and produce 9 *RY* : 3 *RG* : 3 *WY* : 1 *WG*. In Mendel's experiments the actual numbers were 315 *RY* : 108 *RG* : 101 *WY* : 32 *WG*. In the author's repeat experiments the numbers were 997 *RY* : 338 *RG* : 313 *WY* : 107 *WG*. In 1905 the *RY* seeds of one of the F_3 plants of this experiment were sent to the late Mr A. D. Darbishire, who continued

TOMATO

P₁ — *Fireball* X *Golden Queen*
FIERY RED — PALE YELLOW
Red Flesh — *Yellow Flesh*
Yellow Skin — *White Skin*

F₁ — FIERY RED
Red Flesh
Yellow Skin

F₂

9 FIERY RED	3 CARMINE RED	3 DEEP YELLOW	1 PALE YELLOW
Red Flesh	*Red Flesh*	*Yellow Flesh*	*Yellow Flesh*
Yellow Skin	*White Skin*	*Yellow Skin*	*White Skin*

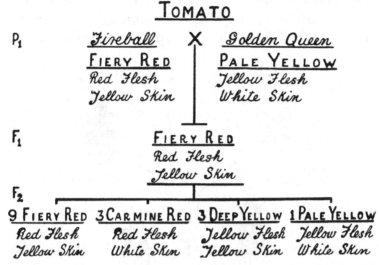

Fig. 10. Showing the compound nature of the red fruit of the "Fireball" Tomato and two new forms, Carmine Red and Deep Yellow, arising after a cross by recombination of the genes.

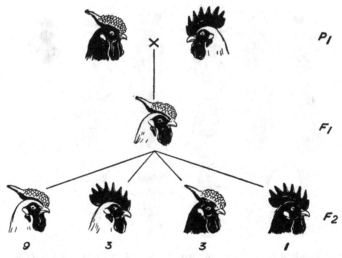

Fig. 11. The assortment of two characters in Fowls, rose and single combs and recessive black and dominant white plumage. The F_1 is white with a rose comb, since these are dominant, but in F_2 we get 9 white with rose combs, 3 white with single combs, 3 black with rose combs and 1 black with single comb, giving the 9 : 3 : 3 : 1 ratio. For comparison, only the male combs are shown. (After Crew.)

the experiments up to the time of his death in 1915. Afterwards the experiments were taken over by Miss Darbishire and Mr Frank Sherlock and carried on to the F_{17} generation in 1918, and the data of generations

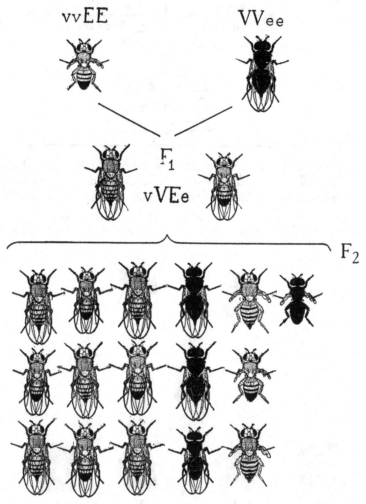

Fig. 12. The independent inheritance of two characters in the Fruit Fly (*Drosophila melanogaster*), long and short wings (long *VV* dominant, short *vv* recessive) and grey and black bodies (grey *EE* dominant, black *ee* recessive), giving the ratio 9 : 3 : 3 : 1 in F_2. (After Morgan.)

$F_{12}...F_{17}$ have been statistically reduced and reported on by Mr Udny Yule in 1923, who found no sensible deviations in the seventeenth generation from the ratios expected on Mendel's law.

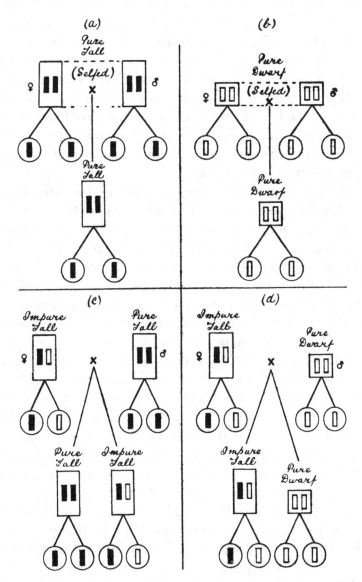

Fig. 13. Diagram showing the results of (a) breeding from the pure tall peas, (b) from the pure dwarfs, (c) back-crossing the impure talls of the F_1 with the tall parent, and (d) with the dwarf parent. The squares and parallelograms represent body-cells carrying the Mendelian pair of genes, while the circles represent their germ-cells carrying the segregated Mendelian genes.

From these experiments we see that since all four possible types occur, seed-shape and seed-colour are distinct characters with an independent inheritance. Other experiments showed that three and even four or more pairs of characters became independently assorted in the germ-cells into which they had entered by a cross. This is known as the Second Law of Mendel, the Law of Independent Assortment of Germinal Units (genes). Mendel's discovery of this elementary fact was one of the secrets of his success, and his experimental demonstration of the existence of single heredity factors proved that *the true unit of heredity is not the individual but the single factor or gene.*

To test his discoveries still further Mendel tried another experiment. He back-crossed the first cross with both parents. Since the first cross has one-half of its germ-cells carrying yellow and one-half green (although it appears to be yellow), when it is back-crossed on to the yellow parent with all pure yellow germ-cells the progeny will all appear to be yellow since yellow is dominant, but on breeding from them it will be found that half of them are pure yellow and the other half impure, the latter carrying the recessive green factor concealed by the dominance of the yellow. On back-crossing the first cross to the green parent the green becomes visible in half the progeny since the 50 per cent. of germ-cells of the hybrid which carry green meet green again, while the other half which carry yellow give offspring which appear to be yellow since the yellow conceals the recessive green. Breeding, however, shows that these are impure yellows. Thus we see the Mendelian expectation once more fulfilled, a 1 : 1 ratio occurring in either case. Back-crossing to the dominant parent, we get one pure to one impure dominant, while back-crossing to the recessive parent we get one impure dominant to one recessive (fig. 13). The pure breeding individuals are said to be *homozygous*, the impure breeders *heterozygous*. Thus in many ways Mendel proved the truth of his discoveries, but he never saw the wonderful fruition of his work. It is pathetic to think of the founder of the Science of Genetics, the Austrian monk, working patiently and alone in the cloister garden at Brünn, where he lived and died almost unknown to the world of science. Like many pioneers in all walks of life, he was too advanced for his own generation, who could see no good in his work and left him to die a sadly disappointed man. For sixteen years after Mendel's death in 1884 the important paper in which he published the results of his experiments with Peas remained unnoticed.

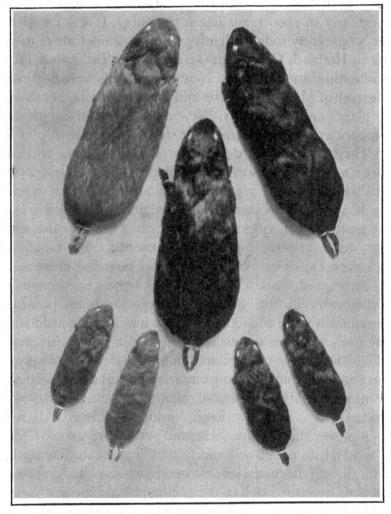

Fig. 14. (Centre) Heterozygous black F_1 Guinea-pig from a cross between a red and a black (above). (Below) Progeny from the black F_1 back-crossed with the red parent (giving 50 per cent. red and 50 per cent. black). (After Castle.)

In the last decade of the nineteenth century a few of us in England and on the Continent were carrying on hybridisation experiments, chiefly inspired by the great naturalist Charles Darwin, to try to find out the laws of heredity, and sometimes we got very near to Mendel's results in our struggles to reach the light. Segregation in F_2 we found in plenty but no gametic purity was established in F_3. The discovery of Mendel's long lost paper early in 1900 reads like a romance. It was found by three botanists, apparently independently, within a short time of each other, De Vries in Holland, Correns in Germany and Tschermak in Austria. A wave of enthusiasm followed. New experiments were begun in order to test the truth of Mendel's theories and to ascertain whether they applied to other plants and to animals. New recruits joined our ranks and a famous school grew up at Cambridge under the leadership of William Bateson. These experiments demonstrated to the full the truth of Mendel's work and every day fresh evidence poured in. Mendel's own experiments with Peas were repeated again and again with precisely the same results, and many other plants were proved to behave in exactly the same manner. The question as to whether the new law applied to animals as well as to plants was also taken up and it was soon found that this was indeed the case. Fowls and rabbits were first dealt with, later pigeons, horses and finally human beings were investigated, though direct experiments in the last case were obviously not possible. This complete confirmation of Mendel's work and the beautiful simplicity of his methods came as a revelation to us all. The great problem of heredity had at last been solved and many set to work to apply it in all directions; to make our food crops more plentiful and of better quality and to improve our livestock and all other plants and animals which are of use to man as a source of food, clothing or beauty. It was soon realised, however, that the problem was not entirely solved. The great mass of material in use showed many complications and some apparent exceptions, though the main law held good throughout. A most pressing problem soon presented itself to the philosophical biologist.

These Mendelian factors, or genes as they are now called—What are they and how are they carried by the germ-cells? One knew by their behaviour when in a recessive condition that they were constant entities, since they preserved their identity intact through many generations of complete domination by another factor, and it was common knowledge that characters suddenly "cropped up", an ancestral character appearing quite unexpectedly after a long period of complete disappearance. The problem was approached from quite a different standpoint, and another branch of biology was called in to our aid—that of Cytology, or the study of the minute cells that go to make up living organisms.

CHAPTER II

CELLS AND CHROMOSOMES

FOR many years it has been known that plants and animals are made up of millions of tiny cells. If we look at a small piece of a plant under the microscope we see something like a honeycomb, and realise that the piece of tissue which looks a solid mass to the naked eye is not so at all but is built up of numberless little box-like compartments fitting together like the comb made by the bees and giving the external appearance of solidity so familiar to our eyes. If we take a high-powered microscope and treat our piece of tissue with some suitable stains to make the contents, which are naturally more or less colourless, more visible, we find that these little boxes or cells are not by any means empty. Indeed, the more we look at them and the longer we study them, the greater is our amazement at the complexity of their contents.

The most striking object in a cell when looking at it through a microscope is a more or less round body in the centre called the *nucleus*. If the piece of tissue we are examining comes from a part of a plant which is growing, such as a root tip or a young shoot or a flower bud, we shall see something extraordinary happening. Some of the nuclei appear as round balls with another small dark round body in the centre, which is called the nucleolus, and a fine network of lightly stained material. Many of them, however, will show another phase with numerous small rod-like bodies, which when stained with suitable dyes become darkly coloured and were for this reason termed *chromosomes*. Although the cells of animals are not so regular as those of plants, taking on various shapes and not forming a regular honeycomb-like appearance except in some organs, but lying in a more free state, the nuclei are the same and the chromosomes are remarkably similar in form and texture throughout the Animal and Vegetable Kingdoms. Fig. 15 shows a piece of tissue from the skin of a salamander (in this case the cells are regularly placed) and another from the root tip of an onion. In some cells the nuclei are lying at rest, as it is termed, the individual chromosomes being invisible. After a certain time these resting cells become very large, too large to remain as they are, and they commence to divide and to form two cells.

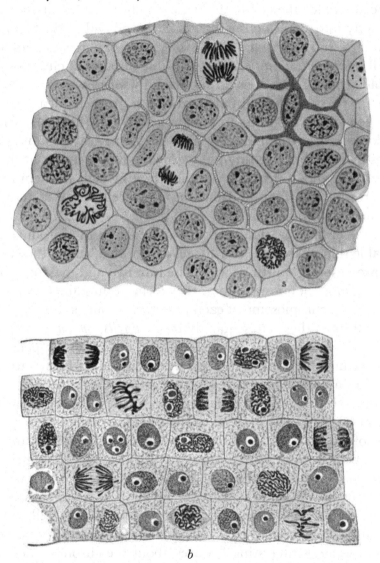

Fig. 15. *a*, a section of the epidermis of a larval Salamander (*Amblystoma*) and, *b*, a section from the root tip of the Onion (*Allium*), showing the cells with their nuclei and the chromosomes appearing, dividing and forming new cells. (After Wilson.)

The chromosomes now begin to appear, and it is soon observed that these are not irregular in number and shape but that each cell contains exactly the same number of chromosomes of similar appearance. When the time comes for the cell to divide, each chromosome divides lengthwise exactly into two, and one-half goes to one end of the cell and the other half to the other end. Thus two groups of chromosomes form in each cell, each containing exactly the same number of chromosomes as in the old cell, and these new groups rapidly form two new cells which are exact replicas of the old one, and which, in their turn, will divide again when grown too large to remain as one cell (fig. 16). When one sees a flower bud growing day by day, the young leaves coming out on the trees in spring, or the tiny baby growing into a strong healthy boy and man, one knows that it is because the tiny cells in them are constantly growing, dividing and making more cells, each containing its full number of chromosomes.

When the animal or plant becomes old enough to reproduce itself a remarkable thing happens. Germ-cells are formed, and in these, instead of the chromosomes splitting lengthwise into two and making cells with the same number of chromosomes, they form up in pairs and reduce so that one whole chromosome of each pair goes to one end of the cell and the other to the other; consequently *the germ-cells contain only one-half of the number of chromosomes present in the body-cells.* Since this happens in both male and female germ-cells, when fertilisation takes place either in a plant or in an animal, the two parental sets of chromosomes come together and the two germ-cells fuse to form a new individual with the original number of chromosomes present in its parents. Fig. 17 gives a diagram of this in an animal and in a plant. The first column down shows the body-cells in male and female with four chromosomes which form two pairs in the next column. These, reducing, give cells with only two chromosomes in each which will function as germ-cells, and the male and female cells coming together in fertilisation will give new individuals each bearing four chromosomes, two from each parent.

One of the most interesting features about the chromosomes is that, although their number and shape are usually constant in each cell of any particular animal and plant and the species to which they belong

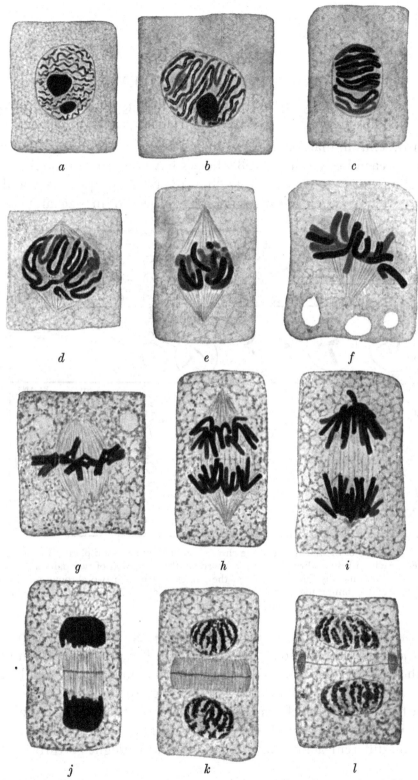

Fig. 16. The division of a cell in the root tips of the Onion (*Allium Cepa*). *a, b, c, d*, the chromosomes appearing; *e, f*, the spindle forming with the chromosomes in the centre; *g*, the chromosomes splitting lengthwise, one-half of each (*h*) going to each pole of the spindle; *i, j, k, l*, the two new nuclei formed each with the same number of chromosomes as the other and as in the original cell. (After Buchner.)

(fig. 18), different species of plants or animals may have different numbers of chromosomes or they may be different in shape or size (fig. 19). Thus human beings have 48 chromosomes of various sizes and shapes in each body-cell (fig. 28), horses have 60, certain monkeys 54, rabbits 44, mice 40, opossum 22, bat 48, peas 14, apples 34, and so on. Roses have their chromosomes less unequal in size and shape

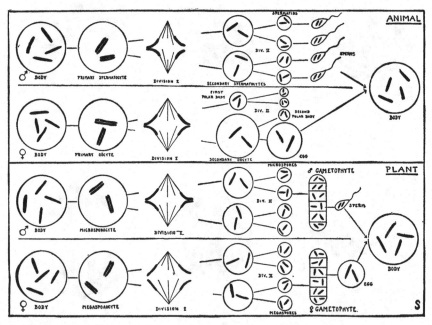

Fig. 17. Diagrams of the behaviour of the chromosomes in animals and plants. To the left the body-cells, each with four chromosomes, followed by the formation of two pairs and the reduction of the number by one-half to form the gametes each with two chromosomes. These (male and female) coming together form the new individuals with four chromosomes in each body-cell. (After Sharp.)

with various numbers in different species (14, 28, 42 and 56), while in lilies there are many various sizes and shapes, but mostly 12 in number, and these variable conditions are found throughout both animals and plants together with variation in number (fig. 19).

The behaviour of the chromosomes was first observed in dead material —that is to say, in material which had been killed and fixed at various stages by different acids and alcohols, so that it might go through the different stages necessary for the preparation of microscopical slides. To do this the material has to go through many chemical processes, to be thoroughly penetrated with paraffin wax and cut in sections perhaps

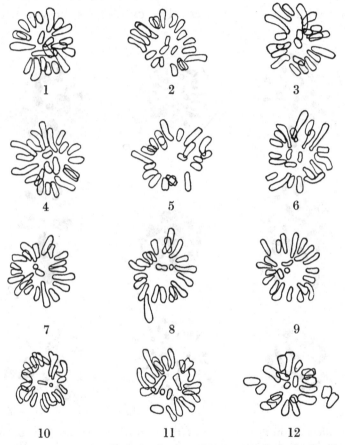

Fig. 18. Somatic chromosome complexes from various parts of the body of the Opossum (*Didelphys virginiana*) (first six from the female, last six from male), showing constancy of size and shape throughout. 1, from the epithelial cell of a renal tubule; 2, connective tissue, cortex of the kidney; 3, lung; 4, epithelial cell of a renal tubule; 5, epithelial cell of intestinal gland; 6, connective tissue of intestine; 7, hepatic cell; 8 and 9, pancreatic cells; 10, cell in thymus; 11, cell in spleen; 12, epithelial cell of adrenal. (After Hoy and George.)

only 1/100 of a millimetre thick, often even less, and then, after being stuck on a slide, subjected to more chemicals and various stains to make the transparent and colourless sections visible. It is not surprising after all these post-mortem processes that sceptics scoffed at the idea of the chromosomes having any reality except in the minds of those who searched for them. It was said that they were artifacts, that is, that they were produced by the action of the chemicals and various processes to which they had been subjected. The cytologists, however, continued

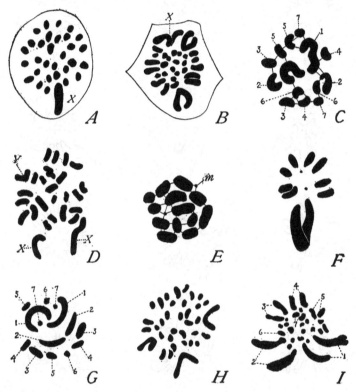

Fig. 19. Size differences in chromosomes. *A*, the Locustid *Orphania denticula*; *B*, *Locusta viridis*; *C*, the Hemipter *Protenor belgica*; *D*, the Mantid *Tenodera superstitiosa*; *E*, the Hemipter *Pachylis gigas*; *F*, the Fly *Drosophila funebris*; *G*, *Aphis rosae*; *H*, the Beetle *Blaps lusitanica*; *I*, from the root tip of a seed plant *Eucomis bicolor*. (After Wilson.)

their work, they felt that their observations must be real or why should the chromosomes appear to go through such regular processes? More-over, identically the same processes occurred in all the many animals and plants which they had examined. It was strange also, if there was no foundation for their existence, that the numbers should be usually identical in every body-cell of any given animal or plant, yet different in animals and plants of other species.

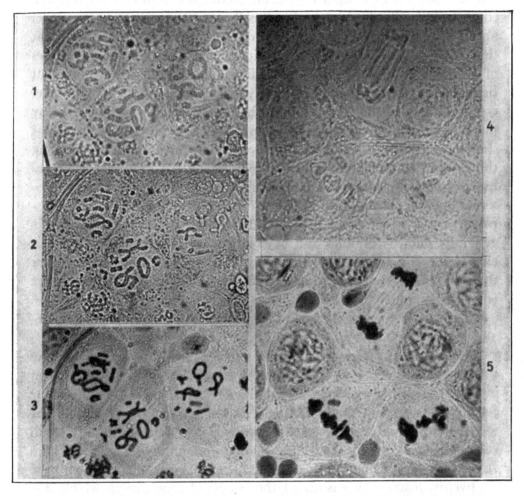

Fig. 20. A comparison of living with fixed and stained material in the Grasshopper *Stenobothrus lineatus*. 1, chromosomes in three living cells about to divide; 2, the same killed and fixed; 3, the same after staining; 4, three more living cells in division, showing one cell nearly divided, the two lots of chromosomes already at opposite ends of the cell, and two cells about to divide; 5, the same fixed and stained. (After Bělař.)

As microscopical technique improved it was found possible to examine these cells in a living condition (fig. 20). By a remarkable culture method pieces of living tissue were kept growing and, as they grew, examined under the microscope. They were put against a dark background with the light falling on them instead of coming through them as is usually the case, and in this way they were thrown into relief and it was possible to follow their every movement. It was soon evident that these cells which looked so stationary on the microscopical slides of fixed material were very much alive indeed. The culture of plant tissue has not been attended with any great success, but animal tissue, especially that of the chick embryo, has been carefully investigated by many observers. In all respects, except that it is alive and not dead, it is similar to the fixed material (fig. 20). There are few more thrilling things than to watch through the microscope the gradual growth of living cells, when, as they get too big for further development, the chromosomes appear, split and divide, thus forming two new cells. The cytoplasm (under which name the contents of the cell around the nucleus is generally known) bubbles and boils and flows this way and that, the chromosomes twist and move from position to position and one realises that the cells, which look so calm and still in the dead material, are in reality brimming over with life and energy, feeding on the supplies extracted from the general food of the animal or plant, extruding waste, and in short behaving much as living individuals themselves.

An interesting fact about "tissue culture", as it is called, is that cultures have been carried on from the same piece of tissue for a great number of years. Provided that nothing occurs to destroy the life of the cells, such as the introduction of poisons, the withdrawal of supplies, or the accumulation of waste products, they are relatively immortal and will continue growing and multiplying indefinitely. Dr Lewis and others in America have cultivated cells from the heart of a chick embryo for twenty years, which is about the length of time during which the technique has been known. Each fragment doubles itself in about 48 hours and is then divided, washed and a fresh culture made with each piece. The growth of these cells by the usual method of chromosome division in the old cells to form two new ones was as rapid after ten years as in the first cultures. Eighteen hundred and sixty generations of sub-cultures had been reached in that time. The late Dr Strangeways

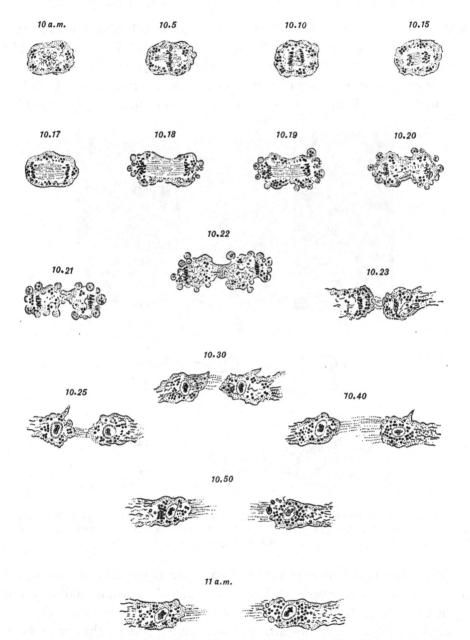

Fig. 21. The division of a living cell in a culture of chick embryo cells. Above are the times at which the drawings were taken, the whole series representing one hour. (After Strangeways.)

had cultures of chick embryo at Cambridge which had been growing for nineteen years, much longer than the average life of the bird. The volume of tissue produced in this time, if left to itself, would be enormous, and one sees the necessity of death and of the reproduction of new individuals to divide up organisms, or they would rapidly become too big

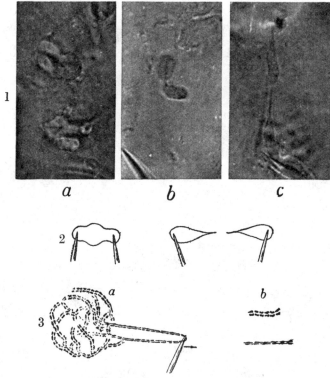

Fig. 22. 1, *a, b, c,* chromosomes in *Tradescantia* (Spiderwort) isolated by needles; 2, line drawing of same; 3, line drawing of chromatic filaments of early stages being stretched with a needle. (After Chambers.)

for this planet. Strangeways found that the cells carried through a new division about every 12 hours, the actual division taking about 35 minutes, though different cells varied from 23 to 65 minutes (fig. 21). Thus the cells are in a constant state of activity, indeed they have been in perpetual motion ever since the first beginnings of life on the earth, since each cell comes from one before and the first cell from a pre-cell back to the first living gene and protogene. In this sense, life may be considered to be immortal, having persisted from the beginning of life and stretching on into the future through the germ-cells and genes, the body being a mere vehicle to act as a protector and food reservoir for the continuance of these perpetuators of the race.

The growth of cells may continue in some cases even after the actual death of the body or what is regarded as personal death. This depends largely on the temperature at which the body is kept, but asphyxiation occurs sooner or later owing to the withdrawal of oxygen supplies. Cells taken from the kidney of a rabbit three or four days after death were successfully cultured and Strangeways actually found cells still living and carrying on divisions in pieces of sausage meat, thus proving that cytological death takes place some time after medical and physiological death.

A brilliant series of investigations on the nature of the chromosomes in a living state has been carried out in America by Prof. Chambers and others as well as on the other contents of cells. The formation and division of the chromosomes in live cells were plainly observed to be taking place in the same manner as in the fixed material. If the membrane of the nucleus was torn the chromosomes fell out or they could be dragged out by means of needles (fig. 22). An isolated chromosome can be stretched with needles and torn into stringy masses, being formed of a more or less gelatinous substance. This practical demonstration of the reality of the chromosomes once and for all dispels any doubt as to the reality of their existence, and it speaks volumes for the rapidity with which technique has improved that these experiments should be possible, since the size of the chromosomes is so minute that it takes a great deal of magnification to make them visible at all.

CHAPTER III

CHROMOSOMES AND GENES

GENERAL breeding experience and genetical experiments agree in showing that the heredity factors or genes can be transmitted equally well by the male plant and animal as by the female. In the process of fertilisation of the female egg by the male sperm, the sole contribution of the male, in most plants and animals, is the nucleus of the gamete. The chief contents of the male nucleus are the chromosomes with their peculiar mechanism and organisation. It is therefore reasonable to infer that the Mendelian factors or genes are borne or carried in or on the chromosomes of the nucleus. As we shall see later, this inference has been amply and completely supported by thousands of experiments. When the chromosomes reduce their number to one-half and form the germ-cells they usually assort themselves at random (with a few exceptions) so that, when the individual is heterozygous by cross-breeding, the chromosomes make various new combinations. Fig. 23 will help to make this clear. Here we have the chromosomes from one parent coloured black and those from the other parent white, a further distinction being the variations in length. The new individual zygote formed by the fertilisation of the egg by the sperm has one pair of each chromosome of the different lengths, one chromosome from each parent. The bottom row shows seven of the various combinations possible in the next generation and it will be observed that the actual number of different combinations that can be formed in this case is no less than 256.

In these chromosomes we find a definite mechanism for the transmission of the Mendelian genes. The segregation of the chromosomes during the formation of the germ-cells provides the random assortment of the genes which was proved experimentally by Mendel.

As a simple illustration of how these recombinations work in a given case we will take the case of two well-known roses that have been crossed in gardens. The female parent of the cross is the wild *Rosa multi-*

flora Thunb. from Japan with tall summer-flowering stems and single white flowers. The male parent is the Old China Rose (*R. chinensis* Jacq.) originally from Central China with dwarf perpetual-flowering

Union of the Haploid Groups. Fertilization.

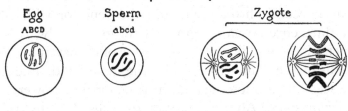

Division of the Diploid Group. Mitosis.

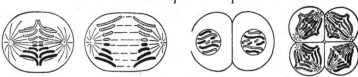

Reduction of the Diploid Groups to Haploid. Meiosis.

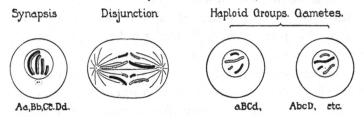

Recombinations in Fertilization.

Fig. 23. Diagram of the life-cycle in animals. 1st row, the union of a sperm- and an egg-cell each with half the number of chromosomes giving the embryo or zygote with the full number of chromosomes. 2nd row, the longitudinal division of the chromosomes giving new body-cells each with the same number of chromosomes. 3rd row, the pairing of the chromosomes and their reduction to form gametes, again with half the chromosome number. Bottom row, showing some of the recombinations possible with four chromosome pairs of different lengths, the chromosomes of each pair being heterozygous for certain genes representing characters (depicted by black or white blocks). Sixteen combinations are thus possible in the gametes, of which only two are shown, and 256 in the zygotes, of which only seven are shown. (After Wilson.)

stems and double pink flowers. Each parent has seven pairs of chromosomes in its body-cells and seven single chromosomes in its reduced germ-cells. From experiments we know that the Mendelian gene for tall summer-flowering stems is located in one chromosome of the seven present in the egg-nucleus of *multiflora*, while the gene for single flowers is located in a second chromosome and the gene for white flowers is located in a third chromosome. Similarly the gene for dwarf perpetual-flowering stems is located in one chromosome of the seven present in the pollen-nucleus of the China rose, while the gene for double flowers is located in a second chromosome and the gene for pink flowers is located in a third chromosome. After fertilisation the cells of the hybrid contain seven pairs of chromosomes, seven from each parent. The chromosome carrying the gene for tall summer-flowering stems from *multiflora* pairs off with the chromosome carrying the gene for dwarf perpetual-flowering from *chinensis* as do the two chromosomes carrying genes for single and double flowers and the two chromosomes carrying genes for white and pink flowers. The resulting F_1 hybrid produces tall summer-flowering stems with semi-double pink flowers, since these are dominant, the dwarf perpetual-flowering stems and single white flowers being recessive. When the germ-cells of the hybrid are formed the seven pairs of chromosomes are reduced to seven singles. Since this reduction is a random one the hybrid produces eight kinds of pollen-cells and eight kinds of egg-cells, so far as the above characters are concerned. Consequently when the hybrid is self-fertilised, in accordance with Mendel's laws, we get, on the average once in sixty-four times, a dwarf perpetual-flowering fully double white rose with the mixed characters of *multiflora* and *chinensis*. Such was the origin of the Polyantha Pompon Rose "Paquerette" raised by M. Guillot in France in 1873, which introduced a new race of roses to our gardens.

An obvious difficulty now presents itself. There are only seven pairs of chromosomes in these roses but there are many more than seven characters, in fact more than one hundred different characters have already been identified in them. How then does one account for this?

It was found at an early stage of the experiments that the genes did not always show an independent assortment. Certain characters, instead of separating in the second generation, remained together, and they were said to be coupled or linked. That one gene was not respon-

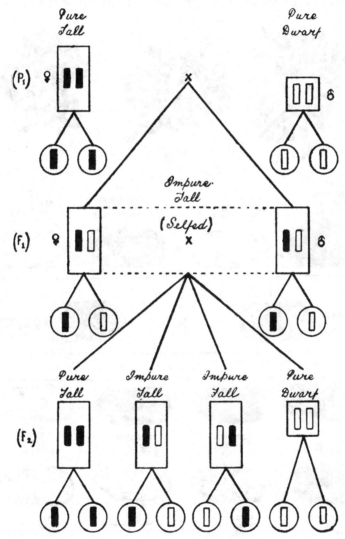

Fig. 24. Diagram showing the segregation of the chromosomes on crossing tall and dwarf Roses; black bars = "tall", white bars = "dwarf" genes in a chromosome.

sible for the whole lot of characters was proved by the fact that at times these characters appeared independently, that is to say, their genes became unlinked. Now if the chromosomes are the bearers of these genes, many genes must be linked together in each chromosome, since there are not enough chromosomes to provide a separate locus for each gene.

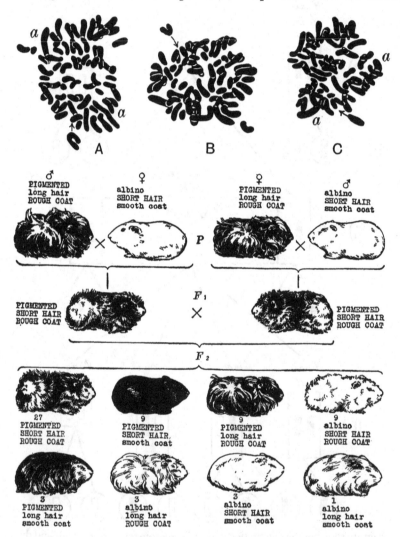

Fig. 25. The assortment of three separate characters in the Guinea-Pig, showing the possible recombinations in F_2, the characters involved being pigmented and albino colour, short and long hair, rough and smooth coats, the dominants being printed in capitals, the recessives in small type. In the three characters segregation of the different types occurs in the ratio of $27 : 9 : 9 : 9 : 3 : 3 : 3 : 1$, only one individual carrying all three recessives occurring out of 64. A, B and C show the somatic chromosomes of the guinea-pig (approximately 62). (After Guyer.)

Here, obviously, was the explanation of the curious fact of linkage which had puzzled various investigators for some time. It now remained to be shown that there were as many groups of genes linked together as there were pairs of chromosomes in the particular plant or animal with which the experiments were being made.

The demonstration of the chromosomes being the bearers of the genes was first experimentally established by Morgan and his col-

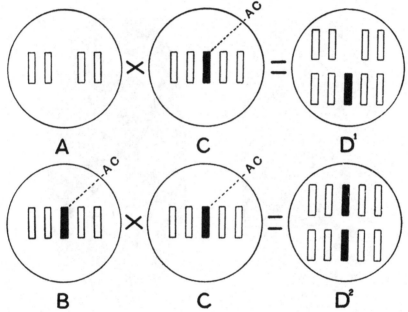

Fig. 26. Diagram showing the difference in the male and female chromosome complex in *Anasa*, in which the females produce only one kind of egg-cell (*C*) but the males two kinds (*A* and *B*), one of which (*A*) has one less chromosome than the female egg-cells. Males (*D*[1]) result from the union of an egg with a sperm (*A*) without the accessory chromosome (*AC*) and females (*D*[2]) from fertilisation by a sperm (*B*) with it. Thus the males contain one chromosome less than the females in their body-cells.

leagues in New York. He had been examining the chromosomes of various flies and found that some of the male and female flies had not identical chromosome groups, one having either one more chromosome than the other or one of a different size. These unlike chromosomes were called sex chromosomes, since their presence in the cells apparently determined the sex of the animal. Various workers had discovered this, and indeed it has been found true for the majority of animals (fig. 26)

and more recently for some of those plants which bear male and female
flowers on separate plants. Fig. 27 shows the fertilisation of the egg of a
Round Worm (*Ancyracanthus*) by two different kinds of sperm. On the
left a sperm with only five chromosomes is joining the egg-nucleus with
six, which will give a male with eleven chromosomes, while on the right
six chromosomes are in the sperm which, joining the egg-nucleus with
six, will give a female with twelve chromosomes. In this worm the

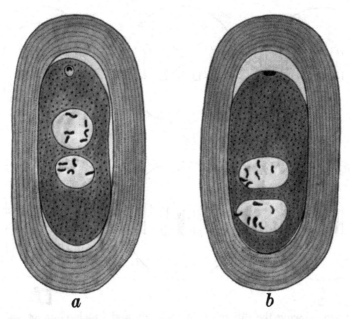

a   *b*

Fig. 27. Fertilisation in the Round Worm (*Ancyracanthus cystidicola*). In each figure the upper
nucleus is the female nucleus with 6 chromosomes, the lower the male nucleus with either 5 or
6 chromosomes. The combination (*a*) 5 + 6 gives a male with 11 chromosomes, the combination (*b*) 6 + 6 gives a female with 12. (After Mulsow.)

female has an extra chromosome while the male has one less in the
body-cells. Consequently in the germ-cells the female always has six
chromosomes in the egg while the male has two kinds of sperms in equal
numbers; in one there are six chromosomes which will produce females
with twelve chromosomes when fused with the six chromosome eggs,
while in the other there are only five chromosomes which will produce
males with eleven chromosomes when fused with the six chromosome

eggs. This explains why males and females are produced in approximately equal numbers.

In those plants which bear male and female flowers on different plants and in some insects and all mammals such as the horse, cow, sheep, pig, rabbit and in man, there are two unequal sex chromosomes, a large chromosome which is called the X chromosome and a small one which is known as the Y chromosome. In all cases so far investigated in mammals the female has two large X chromosomes, while the male has one large X chromosome and one small Y chromosome. In fig. 28 we see the sex chromosomes in man in the act of segregating in the germ-

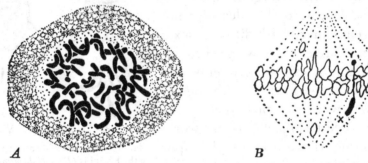

A   B

Fig. 28. Chromosomes in Man (*Homo sapiens*). *A*, somatic chromosomes (48), showing the small *Y* chromosome; *B*, the reduction division, the unequal *X* and *Y* chromosomes shown in black. (After Guyer.)

cells. In the body-cells there are 48 chromosomes and in the reduced germ-cells 24 chromosomes. When the germ-cells are formed the two X chromosomes of the female segregate and every egg-cell carries an X chromosome, but in the male the X and Y chromosomes segregate so that half the sperms carry X and half of them carry Y. The sperms that carry X on fertilising an egg-cell produce female offspring XX, while those that carry Y produce male offspring XY. In the mammals, including man, therefore, it is the male that determines the sex of the offspring, and under normal conditions an equal number of males and females are conceived. In poultry and other birds, however, it is the females which carry the unlike sex chromosomes WZ and the males the like ZZ, so that in this case it is the females that determine the sex. Some butterflies and moths also follow the same scheme as the birds.

while other insects are either like mammals with *XX* and *XY* chromosomes or like some of the thread worms and round worms with *XX* and *XO* chromosomes in which the males determine the sex.

This peculiar mechanism of the sex chromosomes explains many previous difficulties. Many characters are directly associated with one particular sex, for instance, colour blindness occurs mostly in men but it is always inherited through women. Presuming that the genes representing these characters are carried in the sex chromosomes, their explanation becomes quite clear. Hunt Morgan, observing these facts, conceived the idea that not only the genes linked with different sexes but all genes are carried by the chromosomes. Comprehensive experiments were carried out on the Fruit Fly (*Drosophila melanogaster*), which has proved an extraordinarily useful subject for the work, since generations are exceedingly rapid (one a fortnight) and large numbers can be bred in a test-tube with a minimum of trouble, and millions of these flies have been bred experimentally. In 1910 his first results appeared, and a long series of papers culminated in 1925 with a convincing monograph which establishes experimentally the theory that the

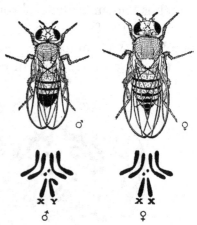

Fig. 29. The male (♂) and female (♀) of the Fruit Fly (*Drosophila melanogaster*) with their respective chromosome complexes. The lower chromosomes in each case are the sex chromosomes, the female having two equal *X* chromosomes, the male having unlike sex chromosomes, an *X* and a *Y*. (After Morgan.)

chromosomes carry the genes. A brilliant series of combined genetical and cytological experiments have not only demonstrated the presence of the genes in the chromosomes, but maps and charts of chromosomes have been made showing their loci or relative positions in the chromosomes. This is one of the most remarkable of the biological achievements of the twentieth century.

CHAPTER IV

MAPPING THE GENES IN THE CHROMOSOMES

A MAP of the genes in the chromosomes of the Fruit Fly (*Drosophila melanogaster*) is shown in fig. 30. It will be seen that there are four groups of genes, and numerous experiments involving many thousands of flies have shown that each of these groups of genes behaves as a unit, demonstrating that the genes within each group are definitely linked. Cytological examination has shown that there are four pairs of chromosomes in the fruit fly, so that each group of genes may be assigned to one of the four chromosomes. These groups are not alike in size, there are two large groups, one medium group and one very tiny group with only a few genes. Similarly, in the chromosomes there are two long chromosomes, one of medium length and one extremely short.

The next question is, how were these maps made showing the actual position of the genes in relation to one another? Also, how do we know that the genes are arranged in a straight line and not in irregular clumps as one might suppose from looking at a chromosome in its later stages of development?

It has been mentioned previously that these linked genes are at times known to become unlinked. This had not troubled the earlier investigators, since in those days the idea of the genes being joined together in any such tangible form as a chromosome was not seriously considered by anyone. Several earlier biologists had an idea of it, especially Weismann at the end of the last century, but as several of his speculations were definitely disproved the part of his work which was more or less true was passed over by the majority of workers.

When Hunt Morgan began his long series of experiments with the flies a large number of characters were studied, and it was soon found that when linked genes did part company it was not a single gene which became separated from its fellows but a whole block of genes. Examina-

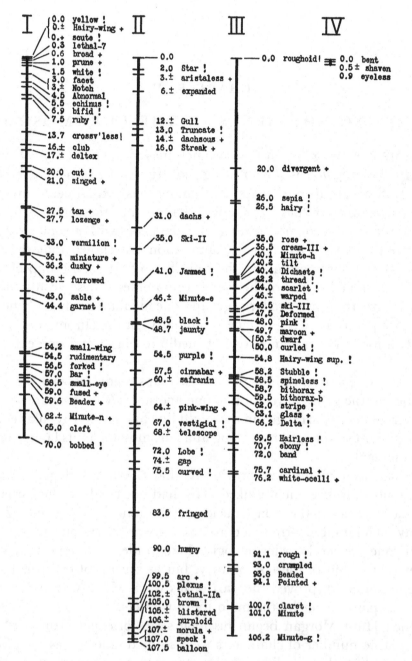

Fig. 30. Maps of the genes located in the four chromosomes of the Fruit Fly (*Drosophila melanogaster*). The numbers stand for the relative distances between. (After Morgan.)

tion of the chromosomes in the affected individuals showed that they appeared to be quite normal, with no evidence of any pieces having broken away. Some time before the chromosomes reduce to form the germ-cells they become extremely elongated and the individuals of the pairs appear to come into close contact with one another at different points. The idea was conceived that, at this time, if the two chromosomes attach themselves too tightly the tension causes them to break and, when they separate, the part which had belonged to one chromosome

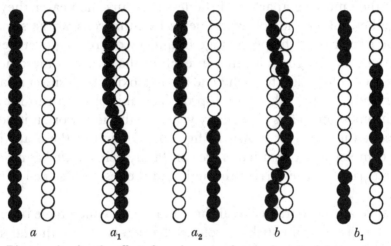

a a_1 a_2 b b_1

Fig. 31. Diagram showing the effect of crossing over of parts of a pair of chromosomes on their gene contents. a_1 and a_2 show a single break; b and b_1 a double one, the black circles standing for one set of linked genes, the white for the other. In both cases new linkage groups are formed. (After Baur.) For simplicity only the two crossing-over threads of the four "tetrad" threads are shown (cf. Darlington, 1931).

had become joined to the other and *vice versa*, with the result that although the two chromosomes of the pair were still identical to all outward appearance, they had really made an exchange of a whole block of genes so that some genes which had been in the one chromosome had crossed over and were now part of the other one and consequently linked up with the other set of genes. Fig. 31 shows a diagram explaining this more clearly, though it must be borne in mind that this is purely diagrammatic. Recent work by Darlington (1931) on this subject is of great interest but of too technical a nature to enter into here.

The idea of the crossing over of parts of homologous chromosomes, though not at the time an observed fact, was a great help in stimulating experiments. Given a linear order of the genes in the chromosomes and the occasional "crossing over" of the genes, as it is termed, it became largely a matter of persistent experimental work to locate the relative positions of the genes in the chromosomes. It is evident that if two pairs of genes lie near together on the chromosome, the chance of crossing over occurring between them is much smaller than if they lie far apart. That is to say, the more crossing over occurring between two genes the farther they must be apart, and the less it occurs the nearer they must be together. Thousands of experiments by numerous workers made it possible to calculate the percentage frequency of crossing over between the various known genes located in each of the four chromosomes and by this means to assign to each gene an approximate position relatively to every other gene, taking an appropriate unit of length to represent each chromosome. Thus after many years of laborious work it is possible to present the maps as shown in fig. 30. Obviously these are by no means complete, new characters are constantly being added as more and more experiments are carried through and the results checked with one another.

A glance at the characters will show that those which have been most freely used are certain striking colour characters or length and size of wings and so forth. Eye colour has proved extremely interesting and afforded much information, not only with regard to the position of the genes, but also as to their behaviour. The main colours—red and white—are sufficiently striking, like the green and yellow peas, to afford a really good clear-cut experiment.

The genes and their linkages have now been studied in many other animals and plants by other investigators. In maize, for example, 180 genes have been worked out and their linkage groups ascertained, and these have been found to conform to the type established for *Drosophila*, showing that animals and plants have identical genetic mechanisms.

With regard to cytological evidence for the presence of the genes in the chromosomes, the chromosomes in most animals and plants are so extremely small that it taxes the highest powers of the most powerful microscopes known at present to examine them in detail, so that until some new technique can be devised, such as the physicists use for the detection of the behaviour of the atoms and electrons, it seems unlikely that we shall be able to identify the actual genes themselves. At present

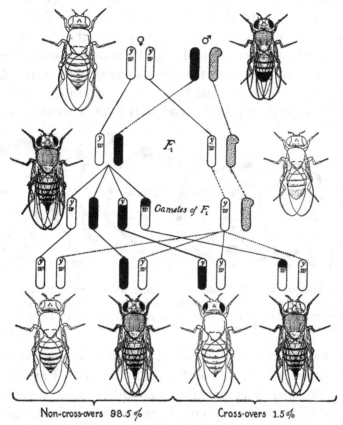

Fig. 32. Diagram to illustrate crossing over in a cross between a female with white eyes and yellow wings and a male with red eyes and grey wings in the Fruit Fly. The blocks represent the chromosomes in which the two characters are linked and the genes are represented by the letters *y* and *w*. In this case crossing over between the genes has taken place in 1·5 per cent. of the offspring, giving flies with red eyes and yellow wings and white eyes and grey wings respectively, instead of the normal linkage in the parents. (After Morgan.)

we have to be content with a knowledge of their behaviour as demonstrated by genetical experiments. These experiments demonstrate that the genes are arranged side by side in linear order on or in the chromosomes, and, since the chromosomes are visible and the genes are not, much can be ascertained about the genes from an examination of the chromosomes which is of vital importance to genetics and evolution. The actual chromosomes which we can see are really the bulky coverings which surround the lines of genes. An examination of certain very large chromosomes at the time of division, notably in *Tradescantia* (Taylor, 1930), shows that the chromosomes consist of a thin outer membrane enclosing a layer of chromosomal sap, in which is contained a long closely coiled thread known as the chromonema, which may be the all important gene line or at least its outer covering. At the actual division the chromosomes are very bulky and it is known by observation and

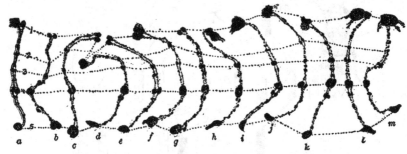

Fig. 33. The chromosome pair "*B*" taken from thirteen different individuals of the Grass-hopper (*Phrynotettix magnus*), showing the remarkable constancy of size and position of the chromomeres during this much elongated condition of the chromosome. The same is true of the same chromosome in different cells of the same individual. (After Wenrich.)

experiment that in some cases at any rate their size is no indication of their gene content, since parts of them may be cast off and left behind with no sensible loss of genic material. Other cases, however, occur where the loss of a small fragment of chromosome or its addition makes a remarkable difference to the characters of the organism. In *Drosophila* the *Y* chromosome, although much longer than the *X* chromosome, is relatively empty of genes contrasted with the *X* chromosome, which carries a very large number of genes. In the early stages before the preparation of the chromosomes for division, the chromonema is extended to its full length and it is at this time that the actual pairing and crossing over are effected. In many cases these stages are too difficult to follow owing to the small size of the nuclei, but some plants and animals have relatively large nuclei and chromosomes in which the processes can be

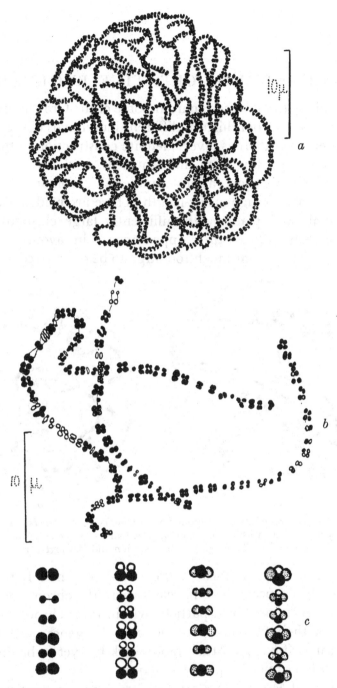

Fig. 34. Examination of the very early stages of gametogenesis of the Lily (*Lilium pardalinum*), when the chromosomes are extended to their greatest length, shows them to be made up of numerous chromomeres. *a*, Upper part of the coils, showing the chromomeres of which the estimated number for the whole cell is 2193. *b*, Portions of three surface coils highly magnified. The chromomeres are seen to be bivalent, each component also in many cases, according to view-point, being double, the smaller units being termed chromioles. They are also seen to be of different sizes and staining capacities. *c*, Diagram of chromomeres and chromioles as seen from different points of view. In this case in the Lily, which is an appropriate subject owing to the very large size of the chromosomes, the whole appearance and behaviour of the chromomeres, including their estimated number, is so close an approximation to that expected for the genes that they may represent actual genes or their outer coverings. (After Belling.)

observed. Fig. 33 shows the *B* chromosome of the Grasshopper (*Phryno-tettix magnus*) taken from thirteen different individuals of the species.

It is evident that there is a very definite construction in this chromosome. Although coming from different cells, a glance will show that each one has definite aggregations of material (chromomeres) at certain marked intervals throughout its length. These are evident at this stage in all animals and plants with sufficiently large chromosomes, and the striking regularity of their appearance is in favour of their being real entities. They seem much too large to be considered as individual

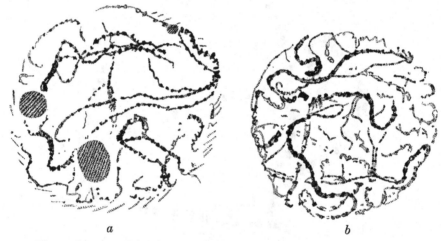

a   *b*

Fig. 35. The conjugation of the chromosomes in pairs in Tulips. *a*, the chromosomes coming together; *b*, lying together with the chromosomes in pairs, in some cases the chromioles are evident. (Cf. fig. 34.) (After Newton and Darlington.)

genes, but they may be linear aggregations of genes. Later work by Belling (1928) on species in Liliaceae, where the chromosomes are exceptionally large, has shown particles within these chromosomes which by reason of their various properties may be genes or at least their outer coverings (fig. 34). Much more work has yet to be done on this subject with finer microscopical technique.

Quoting Hunt Morgan (1927): "We are now in a position to formulate the theory of the gene. The theory states that the characters of the individual are referable to paired elements (genes) in the germinal material that are held together in a definite number of linkage groups;

it states that the members of each pair of genes separate when the germ-cells mature in accordance with Mendel's first law, and in consequence, each germ-cell comes to contain one set only; it states that members belonging to different linkage groups assort independently in accordance with Mendel's second law; it states that an orderly interchange—crossing over—also takes place at times between the elements in corresponding linkage groups; and it states that the frequency of crossing over furnishes evidence of the linear order of the elements in each linkage group and of the relative position of the elements with respect to each other".

CHAPTER V

CHROMOSOMES AND SPECIES

CONSIDERABLE research on the chromosomes of isolated species had already been carried out by cytologists before the genetical importance of the chromosomes was realised, but this discovery has naturally given a great impetus to investigation, and now large numbers of species of plants and animals belonging to many different genera, families, orders and classes have been critically examined and the general constancy of the chromosomes in each species has been definitely demonstrated (fig. 36). Many genera have also been worked out genetically.

One of the most important results of this combined work has been the discovery that the specific rank of an individual can be definitely established by critical and experimental examination of its chromosomes, genes and characters. This method, though somewhat laborious and intensive, has already definitely settled some difficult questions of taxonomy and classification, and will undoubtedly, in the future, be regarded as the only true test of the systematic position of a plant or animal (Hurst, 1925-1931).

By a comparison of the chromosomes and characters of the different species of a genus, and of the genera of a family, order or class, it is possible to elucidate problems of phylogeny and to indicate the most probable course of the evolution and origin of species. Indeed many of the evolutionary processes and mechanisms have already been witnessed in genetical and cytological laboratories and in the cultures of plants and animals used in the experiments. An examination of the chromosomes of various plants and animals shows that there are many different ways by which new species have become differentiated from old species by modifications of their chromosomes, genes and characters. Before dealing with the manner in which these differences have arisen we will briefly outline the various ways in which the species of different genera are found to differ cytologically from one another, since different genera and families as a rule have distinct methods of species formation, and only rarely depart from their own peculiar ways.

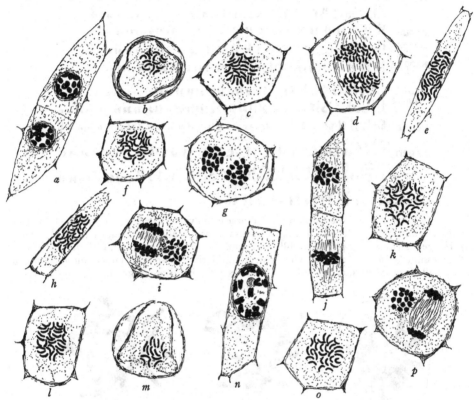

Fig. 36. Chromosomes of sixteen different varieties of one species (*Rosa pendulina* L.) collected by the author from five different localities in the Swiss Alps, illustrating the constancy of chromosome number within a species. *a, h, o, p,* from Goeschenen, Canton Uri; *b, e, f, g, j, n,* from Schuls, Lower Engadine; *c,* from Vitznau, Lake Lucerne; *d, k, l, m,* from Fusio, Canton Ticino; *i,* from St Moritz, Upper Engadine. *c, d, e, f, h, k, l, o,* are somatic cells with 28 chromosomes each taken from leaves, flower buds and achene walls; *a, j, n* are divisions forming the female gametes, with 14 pairs or 14 single chromosomes (the reduced number); *g, i, p,* similar stages in the formation of the male gametes with 14 chromosomes; *b, m,* a later stage showing the 14 chromosomes in the pollen grains.

Comparison of the sets of chromosomes of different groups of plants and animals at corresponding stages often shows wide differences in number, size and shape of the individual chromosomes. Numbers, for instance, may range from one pair only in the Nematode Worm (*Ascaris*)

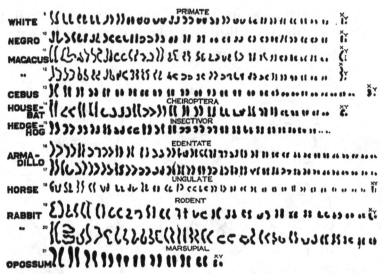

Fig. 37. A comparison of the chromosome complexes in various genera and orders of animals: White men and Negroes (*Homo*, 48), the Monkeys (*Macacus*, 48, and *Cebus*, 54), Bat (*Vesperugo*, 48), Hedgehog (*Erinaceus*, 48), Armadillo (*Dasypus*, 60), Horse (*Equus*, 60), Rabbits (*Lepus*, 44), and Opossum (*Didelphys*, 22), showing general similarity, those with larger numbers having more small chromosomes, those with small numbers having more large ones. (After Painter.)

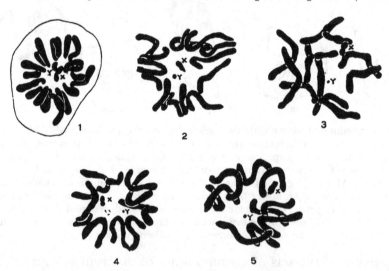

Fig. 38. A comparison of the chromosome complexes of five distinct genera of Marsupials (pouched animals): 1, the Opossum with 22 (*Didelphys*); 2, the Koala or Tree Bear with 16 (*Phascolarctus*); 3, the Tasmanian Devil with 14 (*Sarcophilus*); 4, the Common Dasyure with 14 (*Dasyurus*); 5, the Kangaroo with 12 (*Macropus*). The *X* and *Y* sex chromosomes are marked in each case, the male being digametic for sex. (After Painter, Greenwood, and Agar.)

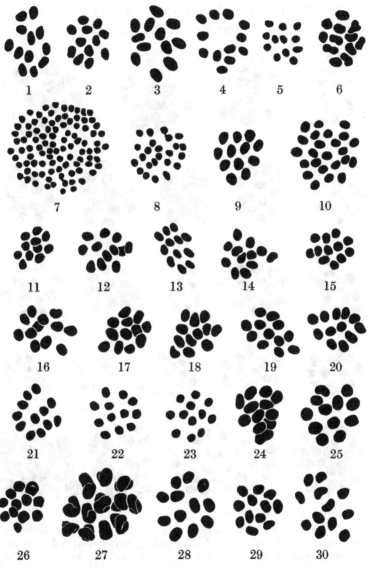

Fig. 39. Twenty-nine species of *Silene* (the Catchfly). 1, *S. maritima*; 2, *S. inflata*; 3, *S. fimbriata*; 4, *S. conica*; 5, *S. dichotoma*; 6, *S. glauca*; 7, *S. ciliata* (Kew); 8, *S. ciliata* (La Linnaea); 9, *S. mentagensis*; 10, *S. vallesia*; 11, *S. saxifraga*; 12, *S. acaulis*; 13, *S. rupestris*; 14, *S. compacta*; 15, *S. nicaensis*; 16, *S. fuscata*; 17, *S. mekinensis*; 18, *S. cretica*; 19, *S. antirrhina*; 20, *S. muscipula*; 21, *S. Behen*; 22, *S. linicola*; 23, *S. echinata*; 24, *S. Friwaldskyana*; 25, *S. tenuis*; 26, *S. Otites*; 27, *S. sinowatsoni*; 28, *S. viridiflora*; 29, *S. nutans*; 30, *s. italica*. All these have 12 pairs of chromosomes except the two varieties of *S. ciliata*, one of which is 16-ploid with 96 pairs of chromosomes, the other tetraploid with 24. A diploid variety also occurs with the normal 12. (After Blackburn.)

to more than a hundred pairs in the Crayfish (*Cambarus*), while in plants, algae and fungi the numbers range from two pairs in Bangiaceae

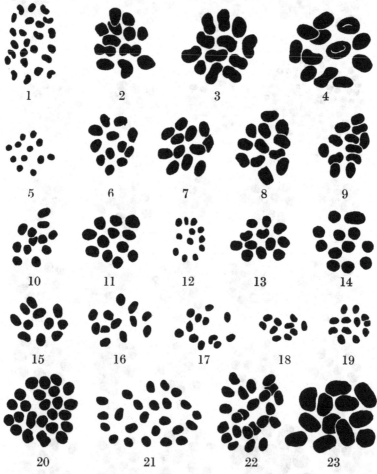

Fig. 40. Twenty-three species of seven genera of the Silenoideae. 1, *Agrostemma Githago*; 2, *Lychnis chalcedonia*; 3, *L. Arkwrightii*; 4, *L. Haageana*; 5, *L. coeli-rosa*; 6, *L. Flos-Jovis*; 7, *L. hybrida*; 8, *L. coronaria*; 9, *L. flos-cuculi*; 10, *Viscaria Sartori*; 11, *V. alpina*; 12, *Petrocoptis Lagascae*; 13, *Heliosperma Alpestre*; 14, *Melandrium dioicum* (hybrid); 15, *M. "yunnanense"*; 16, *M. noctiflorum*; 17, *M. auriculatum*; 18, *M. Zawadskii*; 19, *M. Elizabethae*; 20, *M. virginicum*; 21, *M. californicum*; 22, *M. pennsylvanicum*; 23, *Cucubalus baccifer*. All these have 12 or 24 pairs of chromosomes. (After Blackburn.)

and the Fungi (Taphrinaceae) to more than a hundred pairs in the Horsetail (*Equisetum*), Ferns (*Ophioglossum* and *Ceratopteris*) and a species of *Rumex*. In all recent cases where large numbers of individuals have been examined by several workers it has been found that the number of chromosomes or chromosome sets is constant and characteristic for each species, except in a few cases where races occur with a larger number due to the transverse division of one or more chromosomes. Closely related species and sometimes genera may also have very similar chromosome complexes (figs. 39, 40, 51); for instance, one pair of peculiarly shaped chromosomes was found in twenty different species of *Sorghum* and in some related genera by Huskins and Smith (1931).

Sizes may vary remarkably, however, some genera having chromosomes of approximately similar size with only slight differences, while in others they may vary from some so small that they are difficult to detect to others which may be relatively fifty times as long. Such is particularly the case in birds and reptiles, the domestic fowl and ducks having given considerable trouble to cytologists owing to the presence of a large number of very small chromosomes as well as several long ones, making accurate counts in these species a difficult matter.

We will briefly examine the chief cytological differences which have been discovered between different species.

Many similar species and in some cases higher groups have similar chromosome complexes. In spite of this they are usually inter-sterile, and genetical experiments show that although the chromosomes appear to be externally the same, the contained genes are different, or in some cases where the species are nearly alike they have similar genes but in relatively different positions in the chromosomes (figs. 30, 42, 54). As we have seen, when the pairs of chromosomes form up to make the germ-cells not only do the corresponding homologous chromosomes derived from each parent pair together but the corresponding genes of these chromosomes come together and, as it were, pair also (figs. 23 and 31). Therefore in those cases in which species have the same genes but in different positions, incompatibility arises at once when they form their germ-cells, since when the chromosomes should pair together the genes do not lie in a position to make a perfect pairing possible. Consequently the normal workings of the chromosomes are upset and they fail to produce proper germ-cells or gametes, and the hybrid is consequently sterile. This shows that one cannot judge the relationship of species by examination of the chromosomes alone, such can only be determined by genetical experiments, since it is the genes and not the chromosomes which are of first importance. After the genes the individual chromosomes, each containing a definite complex of genes, rank in importance a good second, while the gametic or germinal set of chromosomes (genome) is, as we shall see later, often of fundamental importance in the determination of species.

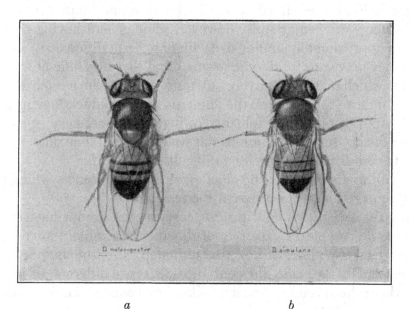

a   *b*

Fig. 41. Males of two similar species of *Drosophila* in which some of the genes have become reversed: *a, D. melanogaster; b, D. simulans.* (Cf. figs. 30, 42, 54.) (After Morgan.)

In other cases we find the same number of chromosomes persisting throughout a large group, but the comparative sizes and shapes of the chromosomes differ from species to species. The best-known example of this is the extraordinary constancy of chromosome number in the genera and species of the Orthopteran Grasshoppers. McClung and his colleagues have examined more than 100 genera, including 800 species of the Shorthorned Grasshoppers (Acrididae), and throughout the family the males as a rule have 23 chromosomes in the body-cells and the

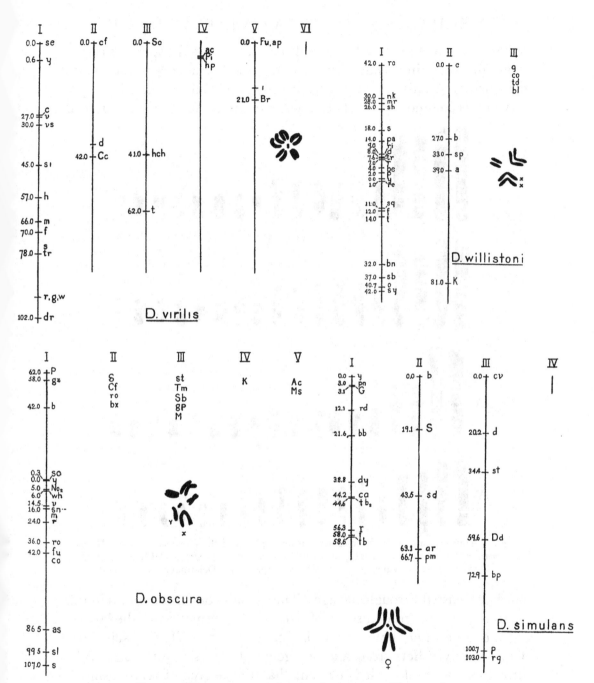

Fig. 42. Chromosome complexes and gene maps of four species
of *Drosophila*. (After Morgan.)

females 24, while the eggs have 12 chromosomes and the sperms 11 or 12. This group of animals has existed for millions of years according to fossil evidence, and only an extremely precise mechanism could have preserved this common series of chromosomes in all the multitudinous cells

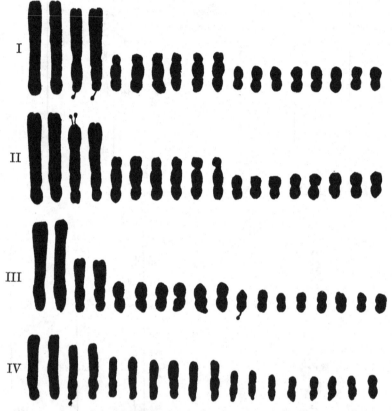

Fig. 43. A comparison of the chromosomes in four species of the Grape Hyacinth (*Muscari*). I, *M. comosum*; II, *M. tenuiflorum*; III, *M. monstrosum*; IV, *M. polyanthum*. All have 18 chromosomes, which vary in form and size. (After Delaunay.)

that have existed through the ages. The 12 pairs of chromosomes in this group represent a graded series of sizes from the smallest to the largest, and the largest may be ten times larger than the smallest. Each genus of this family differs from another genus in the size and form of its chromosomes, and here it is evident that the degree of relationship is as clearly expressed in the chromosome complex as in the external characters of these genera, indicating a descent by modification from a common ancestral series of chromosomes and genes paralleled by corresponding modifications in the bodily structures. A practical demonstration of this was seen when McClung in 1917 discovered a new species of *Memiria* based on a difference in the form of one of the chromosomes. A careful examination of the characters by a systematist showed a

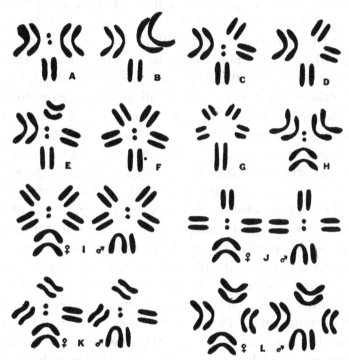

Fig. 44. Different chromosome complexes of 6, 8, 10 and 12 chromosomes found in different species of *Drosophila*. *A* to *H* are female groups, *I* to *L* female (♀) and male (♂) together. In the latter the *Y* chromosome is distinguishable in the male group owing to its different shape and size. (After Morgan.)

corresponding difference in the characters, and the new species was confirmed by him.

Here we have been dealing with a large family in the Animal Kingdom which has presumably preserved the same number of chromosomes through millions of years. A similar case is found in the Vegetable Kingdom among the ancient Gymnosperms including Cycads, Ginkgos

and Conifers. These, so far as examined, have, with few exceptions, each
12 pairs of chromosomes, a condition which must have persisted for
many millions of years.

In other families the chromosome sets or genomes vary in the number
and size of the chromosomes from species to species. This is the case in
species of *Drosophila*, as will be seen in fig. 44. A study of this figure
shows that most of these species may have arisen from a common ancestor by the addition or subtraction of a pair of chromosomes, by fusion
of chromosomes or by the breaking off of parts of chromosomes (fragmentation). Many experiments are being carried out with these species
by various workers to test the relationships between them, but it must
of necessity take many years to get a full and detailed account of the
position of the genes in each, and then to work out the similarity of their
corresponding genes and their relative positions in the different species.

In an analysis of thirty species of these flies Metz has reduced the
different chromosome sets to twelve principal types differing from one
another in number, size and form (fig. 44), and it is evident that the
degree of relationship between these species is equally manifest in the
chromosomes as in the characters, indicating descent from a common
ancestral species by modifications of the chromosomes and the genes
they contain. Similar specific differences in chromosome sets are found
in plants. In the Hawksbeard (*Crepis*), for instance, Navashin made a
comparative study of the chromosomes of species with three, four and
five pairs of chromosomes. Fig. 45 shows how these may be sorted out
according to the shape and size of the chromosomes. The chromosomes
of ten of the species may be sorted out into five different shapes to which
are assigned the letters *A*, *B*, *C*, *D* and *E*. It has been found that the *A*,
C and *D* chromosomes occur in all ten species, the *B* chromosome is in
all but one species, while the *E* chromosome is found in three species
only. The same shaped chromosomes, however, vary much in length
and presumably carry variable numbers and arrangements of genes.

Most genera, in addition to having chromosomes of different numbers
and sizes, show other curious specific differences, some of the chromosomes, or perhaps all, being segmented, usually at the point of attach-
ment of the spindle fibres; that is to say, they are divided transversely

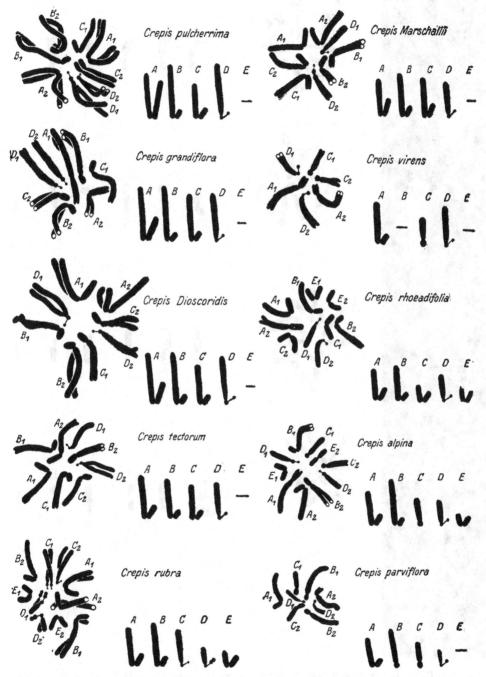

Fig. 45. The somatic chromosome groups in ten species of *Crepis* (the Hawkweed), showing to the right of each a diagrammatic representation of one of each pair of chromosomes lettered respectively *A, B, C, D, E*. A comparison of these shows the variation and similarity between the different chromosomes of the different species. (After Navashin.)

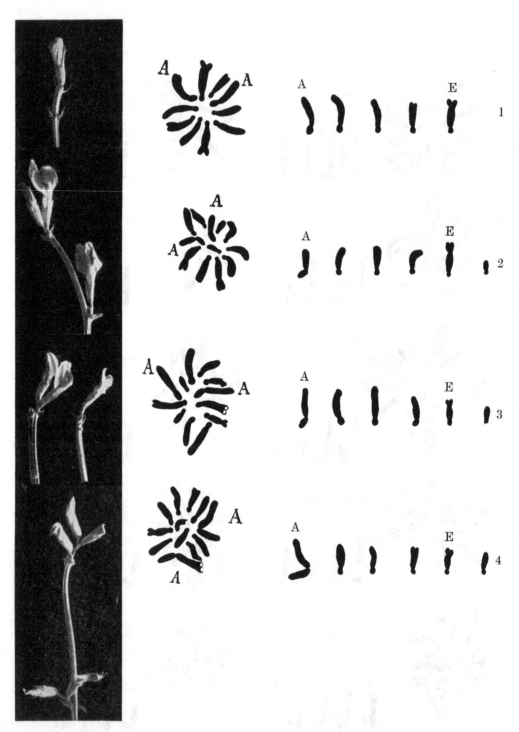

Fig. 46. Different chromosome complexes in species and varieties of the Vetch (*Vicia*). 1, the Spring Vetch or Strangle Tare (*V. lathyroides* L.) with 10 chromosomes; 2, the cultivated Vetch (*V. sativa* L.) with 12 chromosomes; 3, the Wild Vetch (*V. angustifolia* Forst.) with 12 chromosomes; 4, a tall variety of *V. angustifolia* with 12 chromosomes. To the right the individual chromosomes of each set (genome), showing the variation in size, shape and number. (After Sweschnikowa.)

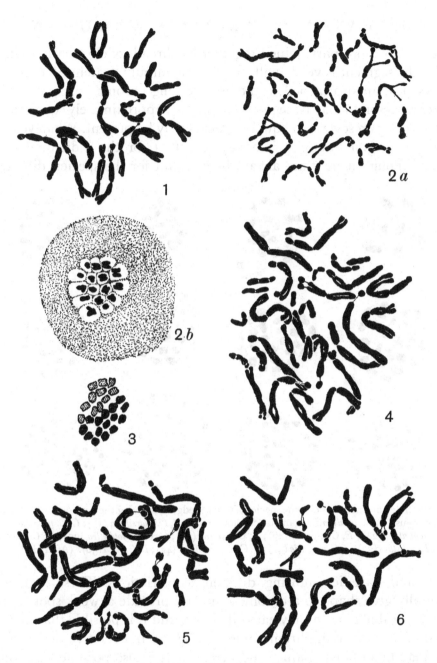

Fig. 47. Chromosomes of species and hybrids in *Godetia*, illustrating segmented chromosomes. A few of the chromosomes show, attached to their ends, the minute portions known as "satellites" or trabants. 1, *G. Bottae* and 2, *G. tenella* produced a solitary highly sterile hybrid which, back-crossed by *G. Bottae*, gave 4, 5, 6, and two other seedlings. (*Magnified* 4,700 *times.*) (After Chittenden.)

into two or more parts joined together by threads of chromatin. Sometimes the segment is very small, such as the minute "satellite" chromosomes (trabants) attached to the ends of long chromosomes, in others the chromosomes may be segmented into two approximately equal parts, while yet again several unequal segments may be present. Whatever the degree of segmentation it is always constant for the individual concerned though it may not always be constant for the species, different

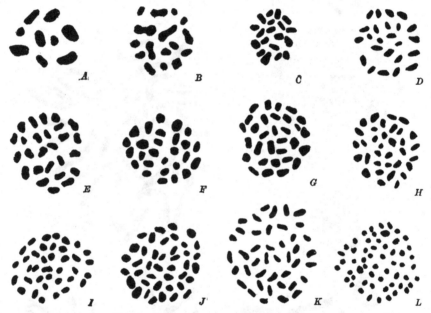

Fig. 48. The chromosomes of some species of *Carex* (Sedges), showing the varied numbers found in this genus. *A, C. pilulifera* (9 gametic chromosomes); *B, C. sparsiflora* (16); *C, C. montana* (19); *D, C. tomentosa* (24); *E, C. digitata* (26); *F, C. loliacea* (27); *G, C. caryophyllea* (29); *H, C. punctata* (24); *I, C. distans* (37); *J, C. vesicaria* (41); *K, C. Goodenoughii* (42); *L, C. hirta* (56). (After Heilborn.)

races and varieties showing different constrictions. These segments naturally give shape to a chromosome and provide a weak spot where a considerable tension may cause division, and it is possible that by this means an increase in chromosome number with a relative decrease in size may be of fairly frequent occurrence. It is also possible that some of the segmentations may mark the fusions of smaller chromosomes, thus giving rise to larger chromosomes from smaller ones, or the junction of part of another chromosome which has become detached from its own chromosome and become joined to another. In any case one would expect frequent crossings over at these weak spots and consequently more exceptions to normal ratios than is usual (fig. 47).

In another old family, that of the Sedges (*Carex*), we find quite a different story from that of the Grasshoppers and Gymnosperms. Here we find a genus with the most variable range of numbers, for out of forty-four species examined by Heilborn twenty-two had different numbers, these varying from 9 to 56 pairs. These varied numbers have evidently arisen by the addition of extra chromosomes and also by fragmentation of large chromosomes or fusion of small ones, since on the whole those species with few chromosomes have large chromosomes while those with many have small ones (fig. 48). It was also found that those species most alike in their external characters were also most alike in their chromosome number and sets, so that it is evident that they were the most nearly related. The nearly allied genus *Scirpus* (Bulrush) has also proved to be similar (cf. also *Iris*, fig. 49).

In many genera of plants recently investigated, such as *Rosa*, *Rubus* (Blackberries and Raspberries), *Solanum* (Potato) (fig. 50) and *Chrysanthemum*, one finds no such irregularity of number but the various species show a range of numbers which are all multiples of one primary number. These are known as polyploid species. Thus in *Rosa* the basic number is 7 and we get a series of multiples of this number in the different species and hybrids, these having in their body-cells 14, 21, 35, 42 or 56 chromosomes respectively. Those having 7 chromosomes in the gametic cells and 14 in the somatic are known as diploid species (carrying two sets, one from each parent), while those having more are known collectively as polyploid species or varieties. These have arisen from the diploids by the duplication of whole sets of chromosomes (see p. 100), polyploid varieties from a single species and polyploid species from more than one species. Thus we see that species and even genera may have similar sets of chromosomes in external appearance but with different combinations of genes or entirely different genes, while in other species we may have the same number but different shapes and sizes; in others again each species may differ in the number of chromosomes or sets of chromosomes (polyploids), while some may even be different in all four, number, shape, size and constitution.

Much sound and solid work has been done in the past on the species

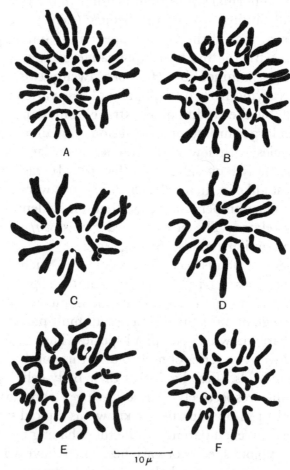

Fig. 49. *Iris* species with different chromosome numbers. *A, I. vaga*, 44 somatic chromosomes; *B, I. xiphioides*, 42; *C, I. Saxi*, 20; *D, I. bucharica*, 22; *E, I. xiphium*, 34; *F, I. juncea*, 32. In *Iris* many species are like *Carex*, the gametic numbers 10, 11, 12, 14, 16, 17, 19, 20, 21, 22, 24, 36, 42 and 56 having been found. (After Simonet.)

problem from a purely taxonomic aspect, that is, by laborious and minute examinations of the external and internal characters of animals and plants. This alone, however, has proved inadequate in many cases,

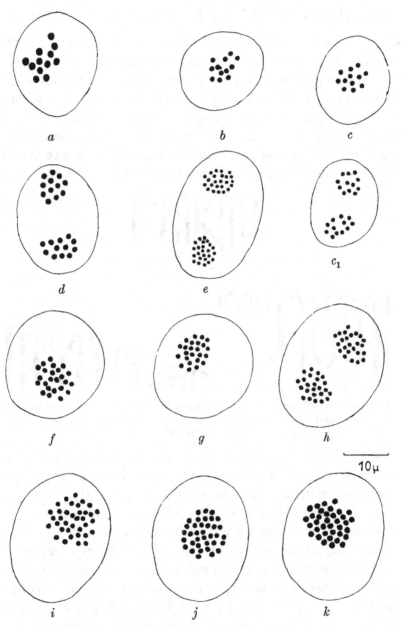

Fig. 50. Gametic chromosomes in *Solanum* diploid and polyploid species. *a, S. sisymbrifolium, b, S. Tomatillo, c, c₁, S. triflorum* and *d, S. Wendlandii,* all with 12 chromosomes (24 somatic); *e, S. laciniatum, f, S. miniatum, g, S. tuberosum* var. "Pepo" and *h, S. villosum,* all with 24 chromosomes (48 somatic); and *i, S. demissum, j, S. guineense* and *k, S. nigrum,* all with 36 chromosomes (72 somatic). (After Vilmorin and Simonet.)

though on the whole this method has been extremely successful and great honour is due to the generations of patient workers who have contributed so much to our knowledge. Many difficulties which appeared almost insurmountable with taxonomic analyses alone are found to be quite simply explained when investigated genetically and cytologically. To work out large genera and families by the combination of all three technical methods is a slow procedure, but it is the only sure and certain way of testing relationships and evolutionary values. A large amount of

Fig. 51. A comparison of the chromosomes of the nearly related genera, 1, *Haworthia cymbiformis*, 2, *Gasteria verrucosa* and 3, *Aloe arborescens*, of the Aloe tribe. The letters mark the probable relationships of the chromosomes. (After Taylor.)

work has already been done on many genera and although only the outskirts of the whole subject have yet been touched it is already possible to present a more definite idea of what a species really is. A species is a group of individuals of common descent, with certain constant specific characters in common which are represented in the nucleus of each cell by constant and characteristic sets of chromosomes carrying homozygous specific genes, causing as a rule intra-fertility and inter-sterility. On this view a species is no longer an arbitrary conception convenient to the systematist, a mere new name or label, but rather a real specific entity which can be experimentally demonstrated genetically and cytologically. Once the true nature of a species is realised and recognised in terms of genes and chromosomes, the way is open to trace its evolution and origin, and the genetical species becomes a measurable and experimental unit of evolution. In many cases the genetical species coincides with the old Linnean species rather than with the new microspecies of Jordan. In some cases, as for instance in the genus *Rosa*, the genetical species proves to be on the whole larger than the Linnean species, corresponding more with the generic section of the taxonomist in the five basic diploid species of the genus, while their sub-species and the polyploid species derived from them by hybridisation in the Pliocene and Pleistocene Periods approximate more to Linnean species.

CHAPTER VI

TRANSLOCATIONS OF GENES

APART from the frequent crossing over between paired homologous chromosomes, many cases have occurred in the cultures of *Drosophila* and other animals and plants in which a piece of a chromosome has become transversely detached and has joined the chromosome of another pair. This gives an entirely new linkage group, since the genes contained in the section are now free from the genes with which they were previously associated and have become part of another linkage group constituted by the chromosome to which they have become attached. In addition to this new association of genes other points of difference arise. The chromosome to which the section is attached becomes longer and the one from which it was broken away becomes shorter than normal. In some cases the lost piece joins on to a chromosome of another pair (non-homologous), but in other cases it joins on to the end of the homologue of the chromosome from which it was detached. When this is the case this particular chromosome will have the section in duplicate while the one from which it was taken will be deficient in these characters. This gives rise to changes in the Mendelian ratios owing to the different balance of genes and also, especially in cases of deficiency, a lethal effect may arise causing the death of the individual (fig. 52).

This changing of chromosome sections is known as "translocation" or "displacement". It is apparent that when a translocation occurs, a deficiency or a duplication is created in the progeny and it is only in later generations that a stable combination will be produced. In those cases in which the translocated section is present in duplicate the appearance of the progeny will be considerably changed, the characters represented by the duplicated section being exaggerated.

The physical basis for translocation is probably explained by the tearing off of parts of chromosomes in the early stages of the divisions of the germ-cells. When the chromosomes are very elongated the two components of a pair come together, often joining up first at the free ends and then coming together along the length of the chromosome. It sometimes happens that part of a chromosome of one pair lies between the chromosomes of another pair. It usually frees itself but occasionally the two chromosomes close in before this is effected and it is torn off. If the chromosome pair should be crossing over at the point of contact the detached section may join on at this point, but it more usually joins on

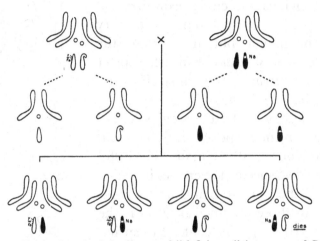

Fig. 52. Diagram illustrating the inheritance of "deficiency" in a race of *Drosophila*. Here experiment has proved, by the unexpected appearance of recessive characters in the progeny of a cross with another race bearing these, that a whole block of genes is missing from the *X* chromosome or they have become inactive. When in a pure state or on meeting the *Y* chromosome this deficiency is lethal. (After Mohr.)

to the end of a chromosome. The detachment may also arise by the failure of the parts to rejoin after a break during the crossing over period or to an accidental crossing over between non-homologous chromosomes. Other ways in which translocations may arise are illustrated in fig. 53.

Besides *Drosophila melanogaster* several other species of this genus of flies have been investigated. Of these perhaps the most interesting is *D. simulans*, which is the only species which will cross with *melanogaster* up to the present. The hybrids are completely sterile but many points of interest have been discovered in them. Externally the two species are very similar, indeed for some time they were not recognised as distinct species and it was only by genetical experiment that the fact was demonstrated. *D. simulans* is smaller and proportionately stouter than *melanogaster* (fig. 41), the wings are relatively shorter, the eyes larger, the external genitalia of the males definitely distinct as well as several minor characteristics of coloration and covering which are obvious on closer examination. Cytologically the chromosomes are very similar in the female but in the male the Y chromosome of *simulans* is much shorter than that of *melanogaster*, being a straight rod about two-thirds the length of the X chromosome, while the Y of the latter species is longer than the X and is the shape of a **J**. Reciprocal matings are not usually equally successful between these two species, the crosses in which *melanogaster* is the mother being much more successful than those in which *simulans* is the mother, from 10 to 40 per cent. of offspring arising against not more than 2 per cent. from the reverse cross. This is due primarily to the dislike of *simulans* females to mating with *melanogaster* males.

A comparison of the genes located in the *simulans* chromosomes with those of *melanogaster* will be found in fig. 54. It will be seen that in chromosome I the two are practically alike, the genes lying in the same sequence. Chromosome II has not been thoroughly investigated, the four discovered loci being in the same sequence but not in the same relative positions. In chromosome III the whole of one section has become inverted as com-

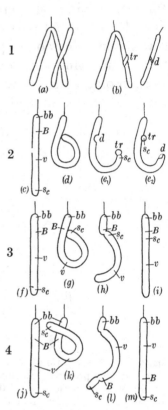

Fig. 53. Diagrams illustrating how different types of chromosome dislocations may occur. 1, translocation of a section from middle of one chromosome to middle of another by non-homologous chromosomes becoming attached and, on breaking away, a piece of one left behind in the other; 2, translocation within the same chromosome (e_1) from centre to end and (e_2) from end to centre; 3, inversion of a part of a chromosome; 4, inversion of the middle section of a chromosome. (After Serebrovsky.)

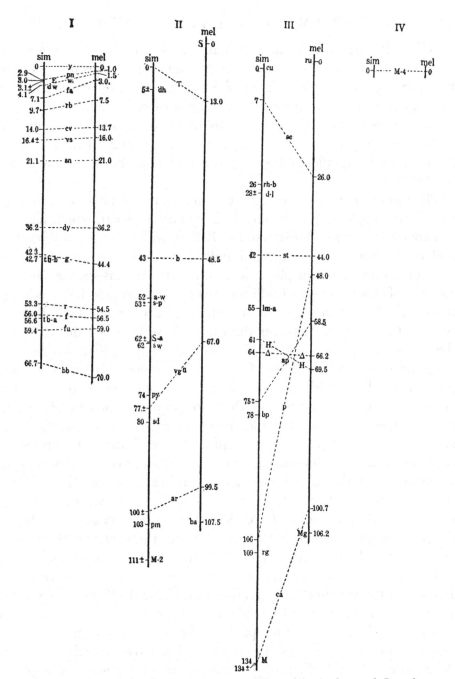

Fig. 54. A comparison of the chromosome maps of *Drosophila simulans* and *D. melanogaster*, showing positions of the same genes in each. Chromosome I is similar in both, II is not well understood in *simulans*, III shows a section inverted. (Cf. figs. 30, 41, 42.) (After Sturtevant, Bridges and Morgan.)

pared with *melanogaster*, and when it was first discovered it was suggested that this inversion might be the cause of the specific difference between the two species. It has since been found, however, that such inversions are not uncommon in *melanogaster* itself within chromosomes II and III without visible bodily differences or occurrence of sterility, so that this hypothesis is no longer probable, though it may be found to have some significance when the other characters are more fully worked out.

The most striking cases of translocations in plants have been found in the Thorn Apple or Jimson Weed (*Datura Stramonium*), which has been intensively investigated by Blakeslee, Belling and others, both genetically and cytologically, thus experimentally demonstrating the transmutations of chromosomes in plants as in animals. In this species there are 12 pairs of chromosomes of different sizes: 5 long, of which one is longer than the others, 5 medium, of which three are longer than the other two, and 2 small, one smaller than the other. They are designated by the formula: $1 L + 4 l + 3 M + 2 m + 1 S + 1 s$. Fig. 55 shows a diagrammatic representation of these chromosomes. It has been found possible to distinguish the two m chromosomes, since one of them possesses a distinct hump, and they are known as m and m^0. In the diagram the L chromosome and the m chromosome without the hump are especially marked and each represented with distinct sections. In (*a*) is shown the condition in the pure line known as Line 1 and in (*b*) the condition in another race known as the "B" race. It was discovered on crossing these two races that certain genes in these two chromosomes were not acting as would be expected in normal cases. Cytological examination of the hybrids also showed that the m chromosomes were joined to the L chromosomes by the ends, thus forming a closed ring of four chromosomes instead of two normal pairs at the reduction division in the germ-cells, suggesting that the ends of these non-homologous chromosomes might be alike in their gene content. Further tests showed that in the "B" race parts of the m and L chromosomes had become interchanged as shown by the distinctive markings of these chromosomes on the diagram, so that on crossing these two together, if we call the Line 1 chromosomes L and m and those of the "B" race

L' and m', we find that in the hybrid the L and L' will join at their like ends (unshaded), but since their other ends will be unlike they will join with the ends of the m and m' chromosomes to which their other parts are now attached, L with the lined end of m', and L' with the lined and stippled end of m, while m and m' will join at their stippled ends which are still homologous. In this way a circle of four chromosomes results

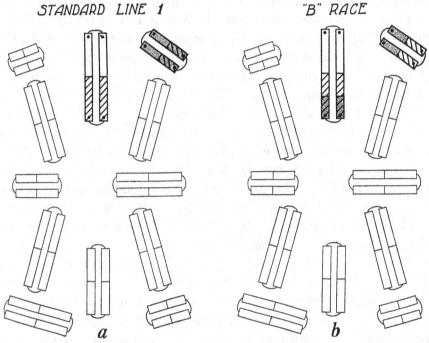

STANDARD LINE 1   "B" RACE

a   b

Fig. 55. *a*, diagrammatic representation of the chromosomes of Standard Line 1 in the *Datura* cultures and, *b*, in the "B" race of the same. The lines and stipples on the L and m chromosomes show the differences of constitution of these chromosomes in the two races caused by the interchange of their parts. (After Blakeslee, *Journal of Heredity*.)

instead of the normal two pairs, and it is obvious that, although the same characters are present in both races and in the hybrid, different linkage groups exist in each.

This interesting occurrence of chromosomal interchange in these two races gave the idea that probably other races might differ in a similar manner, and a new series of experiments was started in which many races procured from various parts of the world were crossed on to Line 1.

The resulting crosses presented many different arrangements of chromosomes, showing that translocations had taken place in many cases between various members of the 12 chromosome pairs. Some had the normal 12 pairs showing no alteration, but from this condition various deviations arose: 10 bivalents (pairs) + 1 circle of 4; 8 bivalents + 2 circles of 4; 10 bivalents + 1 chain of 4; 9 bivalents + 1 circle of 6. The chains arise by the reversal of the segment when the translocation takes place, so that the end is turned inwards and therefore cannot become joined up to the corresponding segment in the pure line. That the Line 1 arrangement is the most usual in the United States is shown by the large number of races from that country which showed normal pairing when crossed with it. The "B" type, however, showing a circle formed by the L and the m chromosomes in the hybrids with Line 1, was the most widely distributed, occurring in races from most of the countries from which they were procured. Not only have the different races within *Datura Stramonium* shown the widespread occurrence of segmental interchange between non-homologous chromosomes, but various species hybrids raised have shown similar aberrations, and it is hoped by a further study of these to be able to throw considerable light on the origin of species in this genus.

An interesting and important case of this chromosome linkage occurs in the genus *Oenothera* (Evening Primrose). A vast amount of work by de Vries, Gates, Renner and others from a genetical point of view showed many extraordinary and unusual features in this genus. In some species it was found that while they breed true themselves, on being crossed they throw two distinct types of offspring, the ordinary Mendelian segregation being absent. An examination of the pollen grains showed that there were usually two distinct kinds formed by these species, large and small, or with different kinds of starch grains, and so on. It was also found that about one-half of the seeds formed were empty. It was therefore concluded that these species were heterozygous, carrying two distinct combinations of characters, and it was only when the two different ones came together that viable seeds were formed, those cases (approximately 50 per cent.) where like met like no seeds arose since they were lethal in a pure condition. Thus these apparently pure-breeding species are in reality permanently heterozygous, the condition being concealed by the death of all zygotes in which both super-linkage groups are not present.

Cytological investigations by Cleland and Hakansson showed the reason for this curious behaviour, the species in which it occurred showing, instead of the normal pairing of chromosomes, a varying number

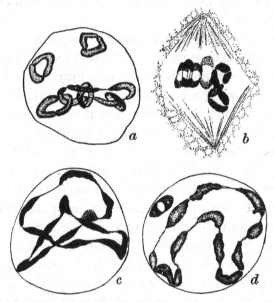

Fig. 56. Chromosomes of *Oenothera*, showing the arrangement of chromosomes in circles and pairs. *a, Oe. franciscana*—circle of 4 and 5 pairs; *b, Oe. rubricalyx*—circle of 6 and 4 pairs (preparing to reduce); *c, Oe. angustissima*—all 14 chromosomes forming a large circle; *d, Oe. Lamarckiana*—a circle of 12 chromosomes and 1 pair. (After Gates, Sheffield and Cleland.)

of ringed pairs and circles, the latter being made up of more than one pair of chromosomes. *Oenothera Lamarckiana*, in which the condition was first found, has only one pair of chromosomes, the other 12 chromosomes (there being 14 in all) forming a large circle. By the arrangement of this circle at the reduction division when the germ-cells are formed, it was seen that they always reduce in such a manner that the same chromosomes go together each time, there being no Mendelian segregation by independent assortment among them. Thus two kinds of germ-cells are always formed, one carrying one complex and one the other. The odd-ringed pair apparently contains only homozygous characters, since it is the only means by which Mendelian segregation could occur. Other species show varying numbers of pairs and circles, from all 14 chromosomes in one circle to those in which 7 pairs are formed in the normal manner. The permanent heterozygotes invariably show a large number of their chromosomes in circle formation, while those which show little or no heterozygosity have no circles but only paired chromosomes.

Several suggestions have been made to explain this phenomenon, but in the light of the recent work on *Datura* it seems that it is caused by the interchange of segments between non-homologous chromosomes, thus giving them more affinity with each other than there is between the original pairs, which have now become unlike by reason of the translocations of the sections. The varying numbers of circles and pairs in different species denote the degree of segmental interchange which has taken place. So far, of the fifteen different arrangements possible twelve have been found and, as shown by Blakeslee and Cleland (1931), from the normal condition of 7 pairs, a comparatively short series of segmental interchanges will produce all these different types, even those with the largest circles. The different translocations which may take place, however, to form these types is of course almost unlimited.

Supposing the original type to be represented by the figures $1.2, 3.4,$ $5.6, 7.8, 9.10, 11.12, 13.14$ (each figure representing one end of a chromosome), if a translocation takes place between the first 2 pairs of chromosomes such as occurred in the two races of *Datura* we get the new chromosome complex $1.4, 3.2, 5.6, 7.8, 9.10, 11.12, 13.14,$ which at the formation of the gametes will show a circle of 4 and 5 pairs. By further interchanges in this manner one can get all the varying types of circles and pairs and it provides an adequate explanation of the evolution of these species. In *Oe. Lamarckiana*, for instance, with 1 pair and a circle of 12, there are two distinct character complexes known as *gaudens* and *velans*. If we apply the same formula as before and

$$velans = 1.2, 3.4, 5.8, 7.6, 9.10, 11.12, 13.14,$$
$$gaudens = 1.2, 3.14, 5.6, 7.4, 12.10, 11.8, 13.9,$$

then when the gametes are formed one pair is formed from the first two

chromosomes and the others from one large ring as follows (one complex
in normal type, the other in italic type):

$$I . 2 \quad 3 . 4 - 4 . 7 - 7 . 6 {-} 6 . 5 - 5 . 8 - 8 . 11$$
$$I . 2 \quad 3 . 14 {-} 14 . 13 {-} 13 . 9 {-} 9 . 10 {-} 10 . 12 {-} 12 . 11$$

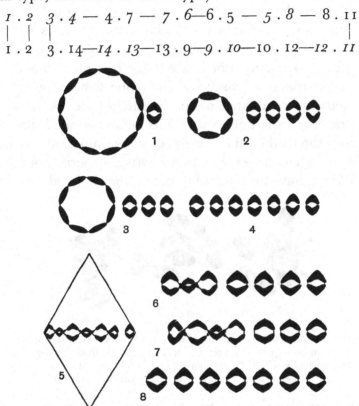

Fig. 57. Diagram showing the peculiar arrangements of the chromosomes in the germ mother
cells of five species of *Oenothera* instead of the regular pairing. 1, *Oe. Lamarckiana* with one large
circle of 12 chromosomes and 1 pair; 2, *Oe. rubricalyx* with a circle of 6 and 4 pairs; 3, *Oe. rubri-
nervis* with a circle of 8 and 3 pairs; 4, *Oe. blandina* and *Oe. deserens*, both of which show the
normal arrangement of 7 pairs; 5, 6, 7, 8 show the respective arrangements of the chromosomes
on the spindle at the reduction division for the formation of the gametes, the chromosomes remaining joined together so that there is no segregation except between the free pairs. (After
Cleland.)

When the actual reduction takes place the circle takes up the position
on the spindle shown in the diagram, fig. 57 (5), and one can see how
the *velans* complex will go into one germ-cell and the *gaudens* into the
other every time except in rare instances when some abnormality occurs.
In this way the two complexes are preserved and owing to the lethal

effect of either of them in a pure state new individuals can only arise when both are present, and by this curious mechanism *Oe. Lamarckiana* appears as a pure-breeding species in spite of the dual nature of its gametes. The most important point about these species undoubtedly lies in the fact that by this unusual chromosome configuration these translocations which, like many others, are lethal in a pure state are able to persist in a heterozygous condition from one generation to another, giving the outward semblance of an absolutely pure species. Only the extensive breeding experiments with this genus revealed the true state of affairs, and the further aid of cytological examination was necessary to provide an adequate explanation. Other genera, notably *Tradescantia* and *Rhoeo*, have been found to bear these circles also and it seems

Fig. 58. Circle of 4 chromosomes occurring in the cross between two races of Peas in which a translocation of chromosome parts has taken place. As will be seen from the figure, these chromosomes may reduce normally *AB–A′B′*, or abnormally *AA′–BB′*, the latter case causing the death of all gametes in which they occur. (After Richardson.)

likely that when more work has been done on other genera their presence may be found to be widespread.

Not only do we get unusual linkage groups in the cases where these circles occur but they may also account for some of the hitherto inexplicable occurrences of sterility. A good example of this has been observed in the Garden Pea (*Pisum sativum*). A cross was made by Miss C. Pellew between an individual belonging to a race of peas in cultivation in Tibet and one of our own edible varieties. Among the descendants sterility occurred in approximately 50 per cent. of the ovules and pollen grains. Cytological examination by Miss Richardson showed that in the affected plants five paired chromosomes and one circle of four are present at the reduction division instead of the normal seven pairs. When the paired chromosomes reduce one chromosome of each pair goes to each pole but in the ring of four it is different. Here we have two pairs AA' and BB' joined together and at the reduction division instead of getting the usual reduction of each pair we may also get the two A chromosomes going to one pole and the two B chromosomes going to the other. When this occurs each of the resulting daughter cells has one chromosome missing from the complete set, their composition being $AA'CDEFG$ and $BB'CDEFG$, respectively, instead of the normal $ABCDEFG$. These will be sterile and since by the laws of chance this should happen in 50 per cent. of the cells it explains the occurrence of the 50 per cent. sterility found in these peas. In *Zea Mays* (Maize), McClintock also found segmental interchange between two non-homologous chromosomes giving rise to a ring of four in the reduction divisions and to 50 per cent. sterility in the gametes.

This random assortment of the components of the circles is not a constant occurrence, however, as we have seen in *Oenothera*, and in the *Datura* experiments it was found that in the hybrids of Line 1 and the "B" race the pollen was good, showing that there was something preventing random assortment in this case. The hypothesis was formulated that at the time of reduction like repels like, the attraction between them only persisting up to this time in the same way that in normal cases the paired chromosomes separate and go away from one another at this period. If this is so the random assortments of the chromosomes in the circles will not occur and the germ-cells will only contain the complexes normal to the two races. In those cases where random assortment takes place the peculiar conditions must have broken down for some reason, the extent of the segmental interchange probably having considerable influence. In hybrids between Line 1 and some other races of *Datura Stramonium* 50 per cent. abortion was observed in the pollen and ovules as in the peas. In a few cases only 25 per cent. aborted, but the reason for this is not yet clear.

That segmental interchange also occurs in the chromosomes of man seems clear, since certain genes have been found to act differently in different families. Thus in some cases the factor for webbed fingers behaves as if it were located in the sex chromosome while in others it appears to be located in another chromosome. That this phenomenon is of widespread occurrence throughout plants and animals is certain, and it must have great significance in the evolution of new species, since

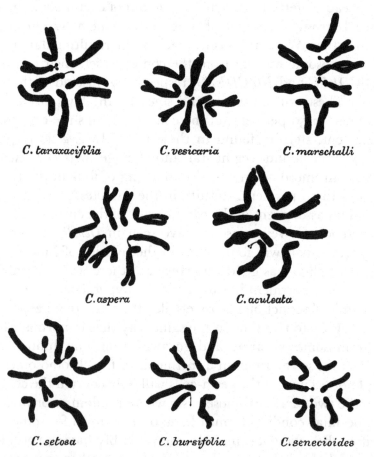

C. taraxacifolia   *C. vesicaria*   *C. marschalli*

C. aspera   *C. aculeata*

C. setosa   *C. bursifolia*   *C. senecioides*

Fig. 59. Figures of the chromosomes of eight species of *Crepis* all having 8 somatic chromosomes. In this genus translocations, duplications and inversions have played a great part in the evolution of the species, since those bearing the same number show much comparative variation in the length and shape of the chromosomes and in the total mass of chromatin. (After Babcock.)

it provides a rapid means of chromosome differentiation which must have a great influence on the evolution of races, varieties, sub-species and species.

This has already been demonstrated in one genus—*Crepis*—which has been investigated extensively by Babcock, Navashin and others from all three standpoints, genetical, cytological and taxonomic. In this genus some species which have apparently the same chromosome complex show a different genetic constitution and it is evident that extensive translocations have occurred between the chromosomes. In other cases the total mass of chromatin differs considerably between the species and there must have been a loss or gain of chromatin during their evolution in the same way that we have seen losses and additions occur in the *Drosophila* cultures. It is also significant in these species that those which have the most similar chromosome complexes are also nearest in their taxonomic relationships. Some cases were also found in which the individuals of one species varied slightly from one another in the length and the shape of their chromosomes, and this may be regarded as a first step towards an evolutionary divergence.

The present wide complexity of the gene contents of the chromosomes must also be due in large measure to the translocations which have occurred in the past. For example, in *Drosophila* the numerous experiments have shown that genes producing reactions which control the same part of the body (such as the eye) may be located in different chromosomes. This may be due to ancient translocations which duplicated the genes for the particular character concerned. Later these would become differentiated from one another by changes in the genes, and we now find these similar genes in the different chromosomes all producing a reaction on the same part of the body. In this way the complexity of organisms must have been immensely increased, and it is evident that translocations giving rise to transmutations of chromosomes have played a great part in the general progress from lower to higher organisms which has taken place and they may be regarded as one of the prime factors in Creative Evolution.

CHAPTER VII

EXTRA CHROMOSOMES

ONE of the most useful results obtained in the experiments with *Drosophila* was the origin of some races of flies which carry an extra chromosome so that one chromosome appears not only paired but in triplicate. By observing which characters were affected by this change it was fairly easy to determine the genes contained in that particular chromosome. This appearance of extra chromosomes has been very fully worked out in the Thorn Apple (*Datura Stramonium*).

In *Datura* the body-cells carry 24 chromosomes (two gametic sets $(2n)$ of 12 each), so that there are twelve possibilities of an extra chromosome appearing. Since the chromosomes are of different sizes in this species (p. 72), it is possible by this means to locate the genes for any particular chromosome by observing first cytologically which chromosome has an extra companion and then genetically which characters are affected. Blakeslee and Belling (1924) have been fortunate enough during a long series of experiments to induce every one of the 12 chromosomes to triplicate in turn so that they now have twelve distinct races each carrying a different chromosome in triplicate. In fig. 60 we see the twelve different effects of the extra chromosome on the seed-capsules. Each of the twelve lower plants has 25 chromosomes $(2n + 1)$, 11 pairs and 1 trisome, as each group of three is termed, since the three like chromosomes are joined together for the reduction division, a different chromosome being trisomic in each case. Several characters of the plant are affected by the change—each type in a different way as a different chromosome appears in triplicate—but here we see the effect only on the seed-capsules, which is, however, sufficiently striking in itself. Above is the normal parent type with 12 pairs of chromosomes $(2n)$. Each trisomic type has been given a name and cultures of them are being raised constantly. Many other interesting types are

Fig. 60. The twelve possible trisomic types in the Thorn Apple (*Datura Stramonium*), showing the variation in the seed-capsules. Above is a capsule from a normal diploid plant. (After Blakeslee.)

being raised from them in which they gain another chromosome or perhaps lose one, so that one gets other types termed $(2n + 2)$ or $(2n + 1 - 1)$ or $(2n + 2 - 1)$ types.

Fig. 61. Capsules and chromosome complexes of extra chromosome transmutant types in *Datura Stramonium*. Above, normal diploid $(2n)$; middle row, left, Echinus mutant $(2n + 1$ Ec$)$ and right, Buckling $(2n + 1$ Bk$)$; while below is the double transmutant Echinus-Buckling $(2n + 1$ Ec $+ 1$ Bk$)$. (After Blakeslee.)

Since these changes are due to chromosome transmutations and not to gene mutations, the types may be distinguished as transmutants. Not only do they get these cases where extra chromosomes occur, and which are known as primary types, but they also get secondary and tertiary types in which the extra chromosome has undergone a translocation.

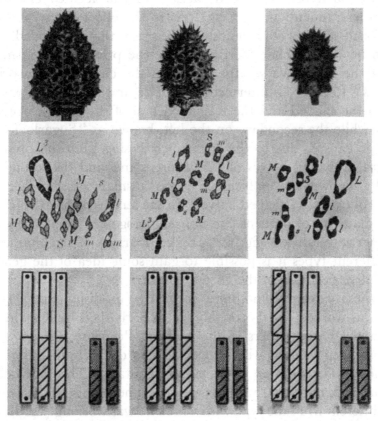

Fig. 62. Primary and secondary transmutant types in *Datura*. In the middle, the capsule, actual chromosome complex and models of *L* and *m* chromosomes (see p. 72) of the primary type known as "Rolled" with an extra *L* chromosome: on the left, those of its secondary "Sugar-loaf" in which the extra chromosome is made up of the unshaded half of the *L* chromosome doubled; on the right, the complementary secondary "Polycarpic" in which the shaded half is doubled in the extra chromosome. In the actual chromosomes it will be observed that in the primary the three *L* chromosomes are joined together at one end only while in the secondaries they form a ring due to the joining of like ends as described on p. 73. (After Blakeslee, *Journal of Heredity*.)

In the secondaries the extra chromosome is made up of one-half of a normal chromosome doubled by a translocation between homologous chromosomes. Fig. 62 shows the appearance of the chromosomes of one of the primaries and its secondary types, in the primary type (middle) the three like chromosomes are joined together at one end, but in the secondary types (right and left) they form a closed ring as we have seen occurs in the hybrids between races with translocated chromosome sections before-mentioned (p. 73). As to the effect of the presence of the different sections of chromosome, it is seen that the secondaries are extreme for certain characteristics while the primary is intermediate. Thus in the figure we see the effect on the capsules. The primary, known as "Rolled", is intermediate in size and shape of capsule as it has both the shaded and the unshaded half of the extra chromosome present, while the secondary on the left, known as "Sugarloaf", has a very large capsule due to the unbalance arising from the presence of two extra unshaded halves of the chromosome, and the secondary on the right, known as "Polycarpic", has a small capsule due to the presence of two extra shaded halves. In the tertiaries the extra chromosome is made up of parts of two non-homologous chromosomes which have been translocated (fig. 63). By comparison of the characters in these different types it is possible to gain some idea of the genes and their relative positions in the chromosomes.

How these extra chromosomes occur by non-disjunction (see *Rosa*, fig. 66) was worked out thoroughly in the early days by Ruggles Gates (1915) and later by others in the Evening Primrose (*Oenothera*). This plant had been worked with extensively by de Vries in the genetical experiments which led up to his famous Mutation Theory (1900). In his cultures there appeared from time to time plants which were distinctly different from the type with which he was working (*Oe. Lamarckiana*), having very broad rounded leaves which were also shorter and more crinkled. The stems were short and irregularly branched, with peculiarly stout buds. The capsules were short and thick, set few seeds and the pollen was almost wholly sterile. An examination of this new type, which was called *Oe. lata*, showed that instead of the 14 chromosomes present in *Lamarckiana* it had one extra, making 15 in all. A great deal of material was examined cytologically and it was discovered that in *Lamarckiana* when the germ-cells were being formed, one chromosome would occasionally go the wrong way during the divisions so that in-

Fig. 63. *a*, primary *m* chromosome type in *Datura* known as "Poinsettia" in which there is an extra *m* chromosome; *b*, secondary type "Dwarf" in which the shaded half of the extra *m* chromosome is doubled; *c*, tertiary type "Wiry" in which the extra *m* chromosome is made up of the unshaded half of the *m* chromosome together with a portion of the shaded half of the *L* chromosome. This constitutes the *m* chromosome of the "B" race (see p. 73). (After Blakeslee, *Journal of Heredity*.)

stead of 7 chromosomes (n) going to each end of the cell one group would have 8 chromosomes ($n + 1$) while the other had only 6 ($n - 1$). Those containing 6 only usually degenerated, not having a complete set or genome of chromosomes, but those with 8, which have a complete set plus one ($n + 1$), continued and made some fertile pollen grains. These, fertilising a normal egg-cell with 7 chromosomes, formed new individuals with 15 chromosomes giving rise to the *lata* transmutant, the changes in many of its characters being caused by the one extra chromosome, present in triplicate, with all the genes it contained, thus upsetting the balance of all the chromosomes. This was the first case of its kind found, but, since, there have been raised in *Oenothera* other types of mutants arising by different chromosomes becoming trisomic, so that they now have seven types in the same way that there are twelve types in *Datura*.

In the Tomato, Schwarzenbach has nine of the possible twelve extra-chromosome types, the tomato having 12 chromosomes in its germ-cells. He also has what he terms secondary trisomic individuals which have only a part of a chromosome extra instead of a whole one. A similar condition was found by Lesley and Frost (1928) in the Ten Week Stock (fig. 64).

Darlington (1928) finds that most of the Sweet Cherries are aneuploid (i.e. have one or more extra chromosomes) and have from 17 to 19 somatic chromosomes instead of the normal 16. These extra chromosomes presumably carry genes which improve the individual from a horticultural point of view, since their occurrence in all the cultivated varieties shows that forms containing them have been unconsciously selected by generations of raisers, who would naturally discard all those which were not an improvement on their general stock. Their continuance in this case is due to the custom of budding or grafting, since seedlings from them often return to the wild diploid state, the odd chromosomes being eliminated in the formation of gametes.

This phenomenon of non-disjunction, as it is called, has been found in many cases in animals and plants, especially in cultivated forms, e.g. in hyacinths, several forms of which have some extra chromosomes. The author has also observed the process in roses (see fig. 66). It is probable

a   b

c   d

Fig. 64. Transmutants in Ten Week Stocks (*Matthiola*). *a*, a very dwarf plant of compact habit which contained a pair of extra chromosomes of very small size, evidently fragments of a larger chromosome, compared with *b*, a normal individual with 7 pairs; *c*, small compact plant of same which contained only one extra fragmentary chromosome and, *d*, another very dwarf plant which proved to have only the haploid number of chromosomes (7 single chromosomes) + 1 fragmentary chromosome. The extra small size of the two very dwarf plants is due to their having the same gene balance, *a*, having 14 normal chromosomes + 2 small, *d* having 7 normal + 1 small, the balance being different in *c*, which has 14 normal + only 1 small. (After Lesley and Frost.)

that, in certain genera where the different species have chromosome numbers which vary slightly from one to another, especially in the more nearly related species, the different numbers have originated in this

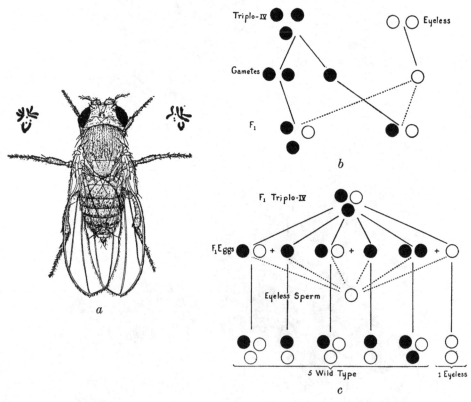

Fig. 65. *a*, a trisomic *Drosophila* (triplo-IV) with 3 small chromosomes instead of the normal 2 (diplo-IV), distinguished by smaller and smoother eyes, darker body, narrower and more pointed wings and absence of trident marking on thorax; *b*, a trisomic (triplo-IV) fly with full eyes mated with a disomic (diplo-IV) "eyeless" fly gives one-half of the offspring triplo-IV and one-half diplo-IV, all with full eyes; *c*, the F_1 triplo-IV with full eyes back-crossed with a diplo-IV "eyeless" gives the usual ratio of 1 triplo-IV : 1 diplo-IV, but there are approximately 5 full-eyed to 1 "eyeless". The same F_1 mated together give about 26 full-eyed to 1 "eyeless". (Cf. figs. 68, 69.) (After Morgan.)

manner. In the Sedge (*Carex*), for instance, in which, as we have already seen, the chromosome numbers are very near to one another but different from species to species, it is highly probable that the numbers have gradually increased, in some cases at all events, through the non-disjunction of a pair of chromosomes in the reduction division. Some recent work with *Viola* by Clausen points to the conclusion that certain species in this genus have also arisen in this way.

The main point is that to form a regular species this odd chromosome must find a mate. If an individual bearing an extra chromosome was self-fertilised and an egg-cell with the extra chromosome was fertilised by a pollen-cell also carrying it, then this regularity would be achieved. As a rule, however, the extra chromosome causes much irregularity in the nuclear divisions and it appears difficult to fix the type. In certain

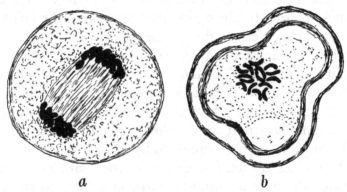

a   *b*

Fig. 66. *a*, a chromosome going the wrong way at the reduction division in a garden variety of the tetraploid rose, *Rosa pendulina* L. (28 chromosomes), both components of one pair going into the same cell so that there would be 15 instead of 14 chromosomes; *b*, a pollen grain of a wild variety of *R. pendulina* found growing in the Swiss Alps, having 15 chromosomes instead of the normal 14, the extra chromosome evidently having arisen by non-disjunction of one pair as in *a*. This phenomenon seems to be much rarer in wild species than in cultivated varieties and hybrids.

Peas (*Pisum*) Håkansson (1931) found a ring of four chromosomes (see p. 78), and in the ovules the entire ring went together at the reduction division giving 50 per cent. sterile ovules and 50 per cent. with two extra chromosomes. If this should also occur in the pollen, offspring would arise with two extra pairs of chromosomes and normal pairing, the dissimilarity arising from the translocations, which previously caused the ring formation, being put right by the duplication of the chromosomes. In such a case the four pairs involved would contain duplicated genes but with different linkages.

That the addition of extra chromosomes may be an important factor in evolution is clearly evident. There are many related genera as well as species which differ by only one pair of chromosomes in number. In some cases this also entails a difference in size which may account for the difference in number, a chromosome having fragmented and formed two, or conversely two may have joined to form one. For example, in the Leguminosae family the Pea (*Pisum*) has a gametic set of 7 chromosomes of more or less equal size, while the Broad Bean (*Vicia Faba*) has a set of 6, one of which, however, is twice the size of the others and has probably arisen from the fusion of two normal chromosomes. Some species of *Vicia* have recently been investigated which have a set of 7 chromosomes apparently arising by segmentation of the large chromosome (fig. 67) or conversely the 6-type has risen from the 7-type by fusion. There are numerous other cases, however, where the chromosomes do not show that such fusion or segmentation has taken place, to which this cannot apply, and it is more reasonable to infer that the difference has arisen by the addition of an extra pair or pairs of chromosomes in the same way that we get the extra one in experimental cultures. A similar case has been observed by Lawrence (1929) in Dahlias. Many dahlia species are tetraploids (see p. 100) with 32 chromosomes, which is twice the normal number, forming usually at the reduction division 8 quadrivalents instead of 16 pairs, showing that there are two of each pair present. One species, *Dahlia Merckii*, has 36 chromosomes and each of the two extra pairs joins one of the quadrivalents making two sexivalents. These two pairs must therefore be homologous with the two quadrivalents and must have arisen by the non-disjunction of two chromosomes or two pairs of chromosomes which became stabilised in later generations. As far as the characters are concerned *D. Merckii* is a very distinct species having the unusual character of fertile ray florets. The presence of the extra chromosomes does not appear to affect its fertility (cf. *Pyrus* and the Pomoideae, p. 145).

That different numbers may arise from the loss of a pair is much less likely, since as a rule at least one complete set of chromosomes (genome) is necessary for the development of an individual. Few cases have yet been found of a transmutant, minus even one of its chromosomes, though it is obvious that when the chromosome goes wandering into

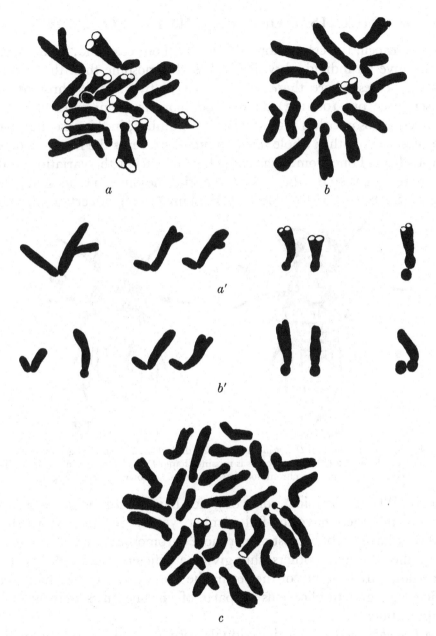

Fig. 67. Two races of the Cow or Tufted Vetch (*Vicia Cracca*, L.). *a*, with 12 chromosomes; *b*, with 14 chromosomes; *a′*, *b′*, a comparison of the chromosomes of the two races showing the origin of the extra pair in the 14-chromosome race by the transverse fragmentation of the first chromosome; *c*, a tetraploid variety (see p. 100) of the same species with 28 chromosomes. (Cf. fig. 46.) (After Sweschnikowa.)

the wrong cell, the cell it should be in is without one, and, if functional, would give an individual with one less chromosome than the normal. Only in cases where there are more than two sets of chromosomes present (see pp. 100–125) do we find more frequent cases of one or more chromosomes less than the normal number; in this case, however, the presence of other whole sets (genomes) prevents the incompleteness of the affected one from having much effect except the variation in the characters caused by the loss of the chromosome and its contained genes. Such a case was found by Huskins (1928) in certain aberrant

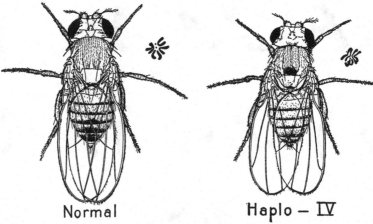

Fig. 68. On the left a normal *Drosophila melanogaster* (diplo-IV), and on the right an individual in which one of the smallest chromosomes is missing (haplo-IV), its loss causing various differences. The respective chromosome groups are also shown. (After Morgan.)

forms of Wheat (speltoids), some of which have 41 chromosomes instead of 42 while others have the full number, yet by their behaviour show that they have probably lost one or more chromosomes, which are made up by the addition of others, thus giving a deficiency and a duplication of certain chromosomes of the set in the same way that we have seen deficiencies and duplications of parts of chromosomes may arise by translocations.

In *Drosophila melanogaster* individuals arise in certain cultures in which one of the small chromosomes IV is missing. These are known as haplo-IV. They are unhealthy, being late in emerging, with a very high mortality, and they are often sterile and always poor producers. They differ in appearance from the normal in several respects, having a paler body colour, large eyes with a rough surface, slender bristles and shorter wings while the aristae are reduced or absent. Haplo-IV also occur in *D. simulans* with similar results. They develop slowly and are nearly always sterile. Individuals have not yet been found in which one of the large chromosomes I, II or III is missing. Presumably the chromosome IV is so small and consequently contains so few genes that its loss is not absolutely fatal, but the loss of one of the large chromosomes

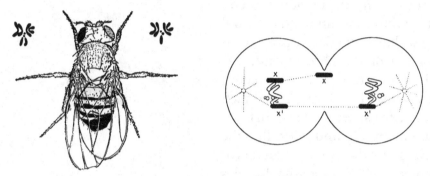

Fig. 69. A gynandromorph in *Drosophila melanogaster*, showing the loss of one *X* chromosome in the right half of the body, making that half male while the left side remains normally female. The left side shows the dominant sex-linked character notched wings (the mother being notched), while in the right side the recessive sex-linked characters scute, broadwing, echinus eye, ruby eye, tan body colour and forked bristles inherited from the father have appeared, showing that the *X* chromosome from the mother parent which contained the genes dominant to these was the one eliminated in the first division of the fertilised egg. (After Morgan, Bridges and Sturtevant.)

would be a very different matter, upsetting the balance of the whole chromosome complex which is essential to perfect development.

Not only do we get chromosomes going the wrong way in the germ-cells but occasionally this happens in the body-cells too. In plants this may give rise to a bud "sport" or bud-transmutation, in which certain characters are different from those of the individual from which it springs. In animals the effect is often even more striking if the sex chromosomes are affected. From time to time individuals are discovered with one half the body male and the other half female, the line of division running down the centre of the body. This is especially striking in various species of birds, butterflies and moths, in which the males have different patterns and colourings from the females. In pheasants, for instance, half the body will be clothed in gay cock plumage while the other half has the subdued feathering of the hen, and in certain butterflies the wings on one side of the body will have the brilliant red, blue or gold colourings common to the male while the wings on the other side are dull brown, grey or white like the females. Not only in external characteristics is the distinction evident, but in internal functionings and sexual characters likewise. An examination of the chromosomes explains these curious and anomalous individuals which have proved such a puzzle to earlier workers. In birds and some moths and butterflies the males carry the two like sex chromosomes (ZZ) while the females carry the unlike (ZW or ZO) (see p. 37). Occasionally in a fertilised egg carrying the two Z chromosomes, which would normally become a male individual, during the first division of the fertilised egg-cell one of the Z chromosomes lags behind, gets lost in the cytoplasm and is consequently missing from the completed nucleus. Thus one of the daughter nuclei contains two Z chromosomes and is male in constitution while the other has only one and is female. Since these two cells give rise by their divisions to the complete body of the adult, one half of the body from the ZZ cell will be male while the other half originating from the cell with only one Z will be female. In some cases the Z chromosome will not be lost during the first division but may be left out in some later division, thus giving

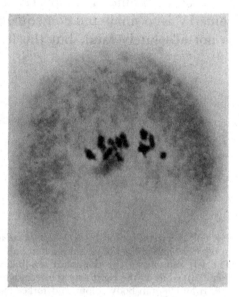

Fig. 70. The loss of a chromosome in a plant of Maize (*Zea mays*). The roots showed the normal number of 20 somatic chromosomes, but examination of the germ-cells showed 9 pairs and 1 single chromosome instead of the expected 10 pairs, showing that 1 chromosome had been left out early in the growth of the upper part of the plant. The plant was small and the tassel poorly developed. (After McClintock, *Journal of Heredity*.)

Fig. 71. A chimera in *Crepis tectorum* in which half the plant (right) has the normal diploid complement of 8 chromosomes while the other half (left) has 9, the *B* chromosome being present in triplicate. Since this plant was one of a culture from a triploid parent it is inferred that the original condition of the plant was trisomic, but during the early cell divisions the extra chromosome was eliminated in one cell which subsequently gave rise to that part of the plant which is normal diploid. (After Navashin.)

rise to female patches on otherwise normal males, the extent of these patches depending on the amount of tissue arising from the cell

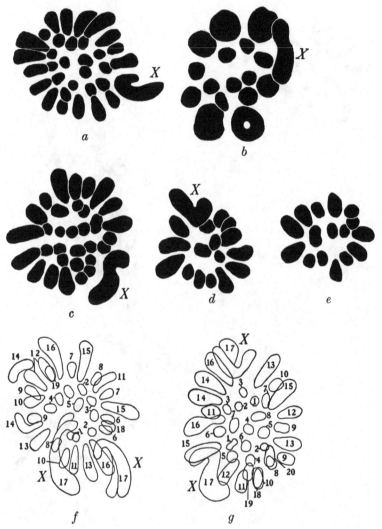

Fig. 72. Chromosomes from a gynandromorph Katydid, *Amblycorypha rotundifolia*. *a, b,* chromosomes from the male tissue of the gynandromorph, somatic and gametic, showing only one *X* chromosome; *c, d, e,* chromosomes of normal male, showing one *X* in the somatic and the two gametic complexes formed, one with and one without the *X* chromosome; *f, g,* chromosomes from the female tissue of the gynandromorph, showing the presence of two *X* chromosomes. (After Pearson.)

originally affected. In animals and certain insects which have the XY type of sex chromosomes (p. 37) we get the reverse happening, male patches occurring in the females, since it is the females which carry the two like chromosomes. These half-male and half-female individuals

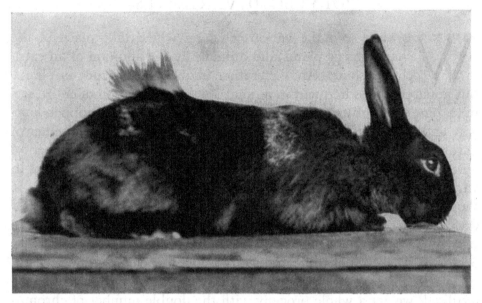

Fig. 73. A Rabbit in which chromosomes have presumably gone astray in the somatic divisions. Its main colour is black and it is short-haired, but it has four patches of brown hair and three of long hair, one of which shows particularly well on its back. This rabbit is heterozygous, carrying genes for brown hair and long hair as recessives, these only appearing in the absence of the dominant genes for black colour and short hair. (After Pickard, *Journal of Heredity*.)

are known as gynandromorphs or gynanders, and of course have no permanent significance except that they provide another substantial proof of the theory that the chromosomes and genes are the organisers and determiners of hereditary characters.

CHAPTER VIII

POLYPLOID VARIETIES

WE have seen the various changes which take place in the characters of plants and animals by the addition of an extra chromosome. Another rather frequent occurrence in plants, but apparently less frequent in animals, is the duplication of entire sets of chromosomes. By some means the division of the cell is stopped half-way and instead of two cells forming, each with the normal number of chromosomes, the original cell remains with double the number of chromosomes. This occurrence has been observed by many workers in various plants and animals in different parts of the body tissue. If it should occur in the first division of the embryo we get an entire individual carrying double the number of chromosomes, since in the subsequent cell divisions the duplicated number persists. Likewise, if it occurs in a part which will produce reproductive elements, we get the germ-cells bearing twice the normal number and in a plant, if self-fertilised, we get a whole progeny with the double number of chromosomes.

This duplication of entire sets of chromosomes, unlike the addition of one or more extra chromosomes, does not usually affect any one character in particular. All the chromosomes being duplicated, all the characters are similarly affected, and the usual result in a diploid species is a plant identical in specific characters with the parent but much enlarged in all its cells and parts except in height. In fig. 74 is an illustration of *Oenothera Lamarckiana* (the Evening Primrose), with its giant variety known as *gigas*. This variety has arisen several times in the cultures of de Vries and other investigators in several species of the genus, and an examination of the chromosomes showed the reason for its gigantism. Each cell, instead of containing the usual 14 chromosomes, has 28, and therefore each gene of the pair is represented four times instead of only twice as in the normal variety.

Fig. 74. On the left, capsules of the diploid *Oenothera Lamarckiana* with 14 somatic chromosomes, and on the right, its tetraploid variety *gigas* with 28 chromosomes. (After Darbishire.)

Fig. 75. *a*, "Telham Beauty", a tetraploid variety of *Campanula* with 32 somatic chromosomes which arose from *b*, the diploid species *C. persicifolia* with 16; *c*, var. *nitida*, a dwarf variety arising by the segregation of recessive genes in various crosses between plants of *C. persicifolia*. (After Gairdner.)

Many of our useful and beautiful garden plants belong to this category, their extra size, compared with the wild forms from which they arose, being the result of their carrying quadruple or tetraploid sets of chromosomes instead of double or diploid sets. Some of the best of our garden roses are of this type, also many flowering bulbs as well as various fruits and vegetables. When the chromosomes are investigated another point comes to light. When the germ-cells are formed the chromosomes, instead of forming up in pairs in the usual way, often form groups of four, the four like chromosomes going together. Fig. 76 d shows this in a duplicated form of the Himalayan rose (*Rosa macrophylla*) known as var. *Korolkowii*. This was found growing in a garden in Khiva and is identical with the parent species except for the extremely large size of its parts. In the illustration of the germ-cells it will be seen that the wild plant with the normal double set of 7 chromosomes (14) forms seven pairs while the garden variety has in this particular cell six groups of four chromosomes each and two pairs, one group of four having failed to come together.

These plants with double the number of chromosomes are called *tetraploids*, showing that they have four sets of chromosomes, in order to distinguish them from the normal plants with two sets which are called *diploids*. A third type may arise from a combination of the other two by cross-breeding. If a gamete from a tetraploid (or an unreduced gamete of the diploid itself) fertilises a gamete from a diploid we get two sets of chromosomes from the one and one set from the other fusing together to make an individual with three sets of chromosomes. This is known as a *triploid*, and many garden plants are of this type, but since they have an odd number of chromosomes they make germ-cells with unequal numbers, which causes the majority of them to be sterile. There are many triploid roses and flowering bulbs, but as the roses are propagated by buds, grafts or cuttings and the bulbs by offsets the question of sterility does not arise, each new plant being merely a piece of the old one. All types which have more than the diploid two sets of chromosomes are known collectively as *polyploids*.

In the numerous cultures of various plants a few plants have arisen which bear only half the number of chromosomes. These are known as

Fig. 76. Tetraploid variety in *Rosa*. On the left, chromosomes, flower and fruit of the diploid species *R. macrophylla* Lindl.; on the right, its tetraploid variety *Korolkowii*. The diploid has, *a*, 14 somatic chromosomes and, *b*, 7 pairs (bivalents) in the germ cells, while the tetraploid has, *c*, 28 somatic chromosomes and, *d*, its germinal chromosomes are often in fours (quadrivalents), the figure showing 6 quadrivalents and 2 bivalents.

Fig. 77. Somatic chromosomes of Tulip species. 1, *Tulipa linifolia* (24 chromosomes); 2, *T. Batalini* (24); 3, *T. Kolpakowskiana* (24); 4, *T. primulina* (24); 5, *T. Orphanidea* (24); 6, *T. sylvestris major* (48); 7, *T. Greigi* (24); 8, *T. armena* (24); 9, *T. galatica* (32). Although there are great differences in the size and form of the chromosomes relatively to one another in different species, the basal number for *Tulipa* is 12 throughout, diploid, tetraploid, and pentaploid varieties having been found. *T. galatica* with 32 (16 gametic) chromosomes was the only exception found, 4 of the chromosomes being very small; the increase is probably due to the transverse fragmentation of 4 larger chromosomes. The related genera *Lilium, Fritillaria* and *Erythronium* also have 12 as the gametic number. (After Newton.)

haploids and are interesting in the fact that they demonstrate that it is possible for an individual to grow which has only one set of chromosomes. Since there are no pairing mates for these chromosomes, the

Fig. 78. Triploid (left) and diploid plants (right) of the Tomato "Dwarf Aristocrat", showing the increase in size with the increase in chromosome number. (After Lesley.)

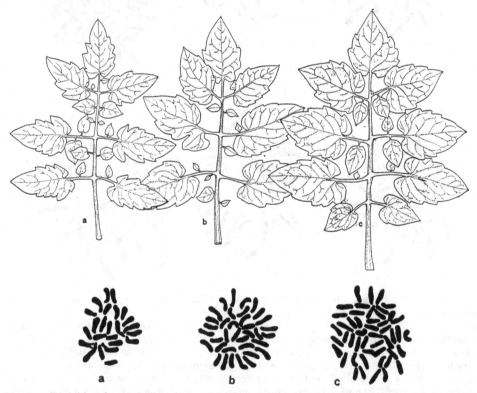

Fig. 79. Diploid and polyploid varieties of the Tomato. *a*, diploid, 24 chromosomes; *b*, triploid, 36 chromosomes; *c*, tetraploid, 48 chromosomes. (After Jörgensen.)

divisions to form the germ-cells are extremely irregular and the plants are usually sterile. In fig. 80 we see all the four types in *Datura*, together with the chromosome sets, diagrammatically represented—haploid,

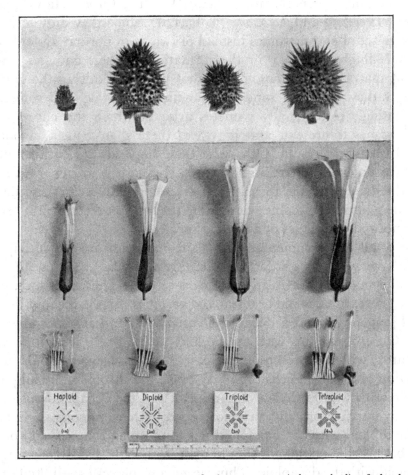

Fig. 80. The capsules, flowers, stamens and chromosomes (schematised) of the haploid, diploid, triploid and tetraploid forms of the Thorn Apple (*Datura Stramonium*). Note the increase in size with the increase of chromosomes. The small size of the haploid and triploid capsules is due to sterility owing to the unbalanced state of the chromosomes. (After Blakeslee.)

diploid (the normal), triploid and tetraploid. It is interesting to observe the gradual increase of the size of the parts with each additional set of chromosomes. The irregular sizes of the capsules are due to the fact that the haploid and triploid forms are sterile and therefore do not produce sufficient seeds to swell out the capsules.

Haploid individuals have occurred occasionally in experimental cultures of other genera—in *Triticum* (Wheat), *Crepis* (Hawksbeard), *Solanum*, *Oenothera* and *Nicotiana* (Tobacco). Since they only have one complete set of chromosomes instead of two as in normal diploids they are exceedingly irregular in the formation of their gametes. Being without mates the chromosomes cannot pair, and during the gametic divisions they assort at random, sometimes splitting and sometimes fragmenting. Occasionally gametes arise in which the entire set of chromosomes is present, and if two of these should meet, a normal diploid plant will result. Otherwise these haploids are entirely sterile and have no evolutionary significance. They are, however, interesting from the demonstration they give that an individual can arise having only one set of chromosomes (providing that one set is complete) instead of the normal double set of a diploid species.

Nearly all the ornamental cherries from Japan are sterile triploids which have arisen by garden culture. Many dahlia varieties are tetraploid and many of the cultivated forms of roses, lilies, tulips, daffodils and hyacinths are triploid or tetraploid varieties. In the Iceland Poppies Miss Ljungdahl found a decaploid variety with 70 chromosomes (ten sets). On crossing this with the diploid 14 (two sets) a regular hexaploid arose with 42 chromosomes (six sets) which formed 21 pairs in the formation of the germ-cells (fig. 83). Since the chromosomes all paired and the cross was fertile it would appear that the decaploid must have arisen by duplications and have contained like chromosomes. By crossing the diploid with the hexaploid fertile tetraploids arose, and on the same basis fertile octoploids should arise with eight sets on crossing the hexaploid with the decaploid. Thus we get a whole series—diploid, tetraploid, hexaploid, octoploid and decaploid—each containing sets of like chromosomes in different numbers. A somewhat similar occurrence was found in the Birches (*Betula*) by Helms and Jörgensen, a natural hybrid being discovered between "diploid" and "tetraploid" forms which showed as a rule 21 paired chromosomes. This number suggests that the "diploid" with 28 chromosomes is really tetraploid and the

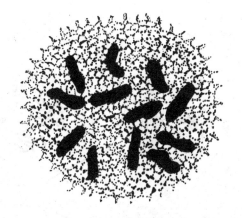

Fig. 81. (Left) haploid Tomato plant, together with its somatic chromosomes (12) above, compared with the normal diploid plant (right), which has 24 somatic chromosomes. (After Lindstrom, *Journal of Heredity.*)

Fig. 82. *a*, haploid (36 chromosomes); *b*, diploid (72); *c*, triploid (108); *d*, tetraploid (144) varieties of *Solanum nigrum* (Black Nightshade). Note the increase in cell size with increase of chromosome number. From the behaviour of the chromosomes in the gametic cells of the haploid, which resembles that of triploids in other plants, one may infer that although 72 is the somatic number for this species it is really made up of six sets of 12 chromosomes (12 being the basal number in the genus *Solanum*) and that it is consequently a hexaploid.

(After Jørgensen.)

"tetraploid" with 56 chromosomes an octoploid, their crossing producing a more or less regular hexaploid with 7 as the basic number of

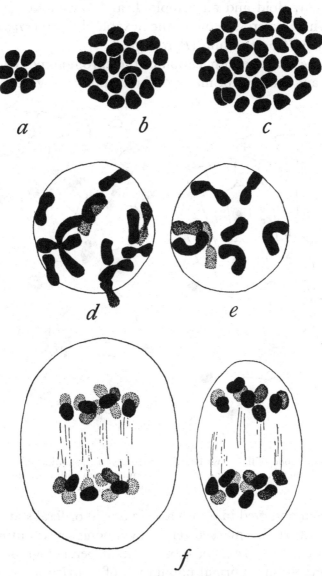

Fig. 83. Polyploidy in the Iceland Poppy. *a*, gametic chromosomes (7) of the diploid species
Papaver nudicaule, *c*, of var. *striatocarpum*, a decaploid with 35 (70 somatic) and, *b*, the regular
hexaploid cross between them with 42 somatic and 21 gametic; *d*, *e*, *f*, gametic divisions in
the crosses, showing 21 pairs and regular divisions. (After Ljungdahl.)

the genus (cf. Woodworth (1929)). In the Foxglove (*Digitalis*) Haase-Bessell (1921) found complete pairing, as in a hexaploid, in a cross between a tetraploid and an octoploid, and Darlington (1927) found regular pairing in a cross between the hexaploid Plum (*Prunus domestica*) and the diploid Cherry Plum (*P. cerasifera*).

Thus we see that the duplication of the chromosomes has been produced in many cultures of many different plants in gardens and they

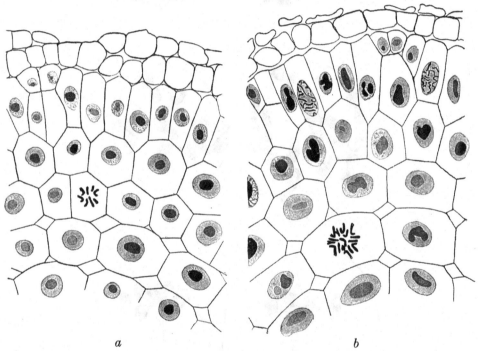

a   *b*

Fig. 84. Sections of roots of the Hawksbeard (*Crepis Bureniana*), showing, *a*, the normal diploid condition with 8 somatic chromosomes and, *b*, a tetraploid area with twice the normal number of chromosomes. (After Hollingshead.)

have also been observed in some cases in a wild or feral state. In *Crepis*, for instance, there is a marked tendency among several diploid species to show with their normal progeny a small percentage of polyploids, Navashin finding in a normal population of *C. tectorum* 0·5 per cent. of triploids and tetraploids. Several external factors may be the cause or occasion of this duplication of chromosomes. A severe frost, for example, will temporarily suspend divisions in the pollen grains and cause some germ-cells to form (provided the frost has not been too severe), bearing twice the normal number. These on fertilising normal egg-cells will produce triploids or, if the egg-cells have been similarly affected, tetraploids. In gardens where many plants are out of their natural environment, having come from countries with more regular climates, such occurrences are not infrequent. In their own country, once the winter is over, they produce their flowers with no set-backs, but in England, where we get a warm spell in the spring long enough to bring out the flowers, often followed by a severe frost for several nights just as the flowers are forming, many aberrations and abnormalities arise.

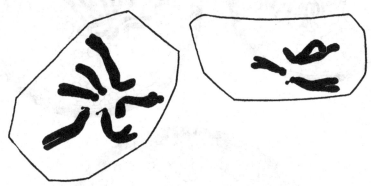

Fig. 85. Chromosomes in the body-cells of the Hawksbeard (*Crepis capillaris*), with a double set of three chromosomes (6) and of its haploid form with a single set. (After Hollingshead.)

Similarly in a wild state, an unusual season may upset many of the normal mechanisms of the plant. De Mol has discovered that many of the different chromosome types in bulbous plants have been due to the custom of drying off the bulbs after the flowering season and, in the case of bulbs for commerce, the subsequent forcing to produce early flowering for culture in pots. These bulbs, being subjected to various degrees of temperature at the time of the formation of their germ-cells (these being formed deep down in the bud during the previous summer or autumn before flowering), produce various irregularities of division (fig. 86). Many of these irregularities fail to carry on but the few that struggle through will give rise to new races, sometimes of great beauty, for our gardens. In wheat plants subjected to high and low temperatures at the time of pollen formation Bleier found many grains with twice or even four times the normal number of chromosomes. Gametes with twice the number were also produced by Rybin and Eghis in *Nicotiana*

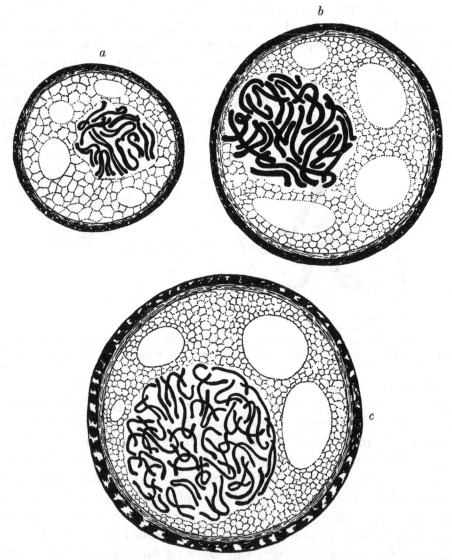

Fig. 86. Polyploid gametes in the Hyacinth variety "White Duc Maxima". *a*, normal pollen grain bearing 12 chromosomes; *b*, diploid grain with 24 chromosomes; *c*, tetraploid grain with 48 chromosomes. (After de Mol.)

(Tobacco) by chloroforming the shoot and thus suspending the reduction division. The Marechalls found that tetraploids could be produced in certain mosses by wounding and from these a few octoploids were produced in the same way. Beyond this, however, it was apparently impossible to go. In many plants and animals X-radiation is a powerful cause of the duplication of the chromosomes.

In certain animals polyploid cells are found in some parts of the body in very high multiples. Some plants also have a certain instability of their somatic cells and these frequently show whole batches of cells

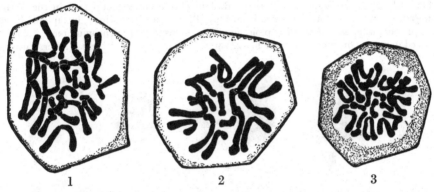

1   2   3

Fig. 87. 1, chromosomes of a triploid Hyacinth "King of the Blues" with 24 chromosomes (12 long, 6 medium and 6 short); and two aneuploid bud variations which arose from it with 18 and 21 chromosomes respectively (2 and 3). Both variants are dwarf, all organs being considerably smaller, and the indigo colour of the "King of the Blues" has changed into carmine in both cases. (After de Mol.)

which have become tetraploid in various parts of the plant. The same phenomenon is found in toads, bees and other animals. In the tomato, especially, much work has been done and it has been found that if normal diploid plants are cut back to a node, a callus forms from which new shoots arise, some of which are often polyploid, usually tetraploid (fig. 88). In this way tetraploid varieties of various races of known genetic constitution have been raised and their study will give invaluable information of the workings of the genes in these polyploids arising from duplication, and probably much may also be learned of the means by which the genes influence characters, since every new angle from which this can be studied must bring fresh light upon the subject. It is

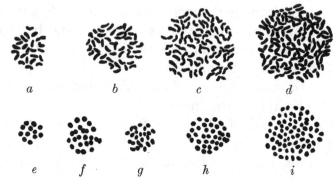

Fig. 88. Polyploidy in *Solanum*. *a*, somatic diploid chromosome group (24) in the Tomato (*S. Lycopersicum*); *b*, tetraploid chromosome group (48) from a *gigas* form; *c*, hexaploid chromosomes (72) of *S. tübingense*; *d*, duodecaploid *gigas* form of last with 144 chromosomes. Below, the reduced gametic chromosomes of each, 12, 24, 24, 36 and 72 respectively. (After Wilson.)

Fig. 89. Capsules and chromosome complexes of six Globe transmutants in *Datura Stramonium*. Above are the normal diploid (2*n*), triploid (3*n*) and tetraploid (4*n*) types. Below are six Globe transmutants, the first two from the diploid with one and two extra Globe chromosomes respectively (2*n* + 1 and 2*n* + 2); the third a triploid with one extra Globe chromosome (3*n* + 1) and the last three tetraploids with one, two and three extra Globe chromosomes (4*n* + 1, 4*n* + 2 and 4*n* + 3). The characters of Globe are stout spines and depressed capsules and these increase with the increase in number of Globe chromosomes. (After Blakeslee.)

also possible to produce the odd chromosome transmutant types from triploid and tetraploid cultures and much work has been done on this in *Datura* (fig. 89).

Winge found that tetraploid cells could be produced in Sugar Beets by inoculating them with *Bacterium tumefaciens*, by wounding or by treating with various chemicals. These cells grew and reproduced very rapidly, giving rise to the cancer-like condition known as crown-gall. The similarity to cancer in animals is particularly striking, since cancer areas are known to have abnormal chromosome constitutions owing to their irregular divisions.

The genetics of polyploid varieties have been worked out to some extent in a few species, e.g. Haldane (1929) in *Primula sinensis*, and the experiments show that new ratios must be expected since each gene of a pair is represented four times instead of twice. We have seen (p. 3) that when we cross two diploids bearing two unlike characters, one dominant the other recessive, we get in the second generation, out of every four individuals, three showing the dominant character and only one the recessive. In the tetraploid, however, where each character of the pair is represented four times instead of twice, we get various complications owing to the occurrence of individuals in which there are three doses of the recessive gene to one dose of the dominant. In some cases the dominant gene is powerful enough to overcome the three recessives entirely, in others, where it is not quite so powerful, we get individuals showing a more or less intermediate condition between the two parental characters, while in others where it is less powerful still the three recessives are able to mask it entirely and individuals arise which show the true recessive character but which carry one dominant gene and are consequently not gametically pure. These will not breed true as we have seen the recessives always do in normal diploid species. Here we get an apparent contradiction of the Mendelian laws of heredity but which further consideration shows to be entirely consonant. Whichever of these three results occurs, the controlling influence apparently being the relative potency of the genes involved, the normal 3 to 1 ratio of the diploids does not occur. In higher polyploids greater complications still arise with each additional pair of chromosomes and genes. As a matter

of fact in most diploids we get cases in which the dominant character
does not entirely conceal the recessive. For instance, in the Four o'Clock
(*Mirabilis jalapa* (fig. 90)) if we cross a red-flowered race with a white-
flowered one the resulting cross will have pink flowers instead of red.
At first sight this looked like a disproof of the Mendelian law of the
purity of the genes, since it is what was expected in pre-Mendelian days
when it was believed that the contribution from each parent became
mixed or blended with that of the other, like mixing two different
coloured dyes together, the subsequent descendants getting different

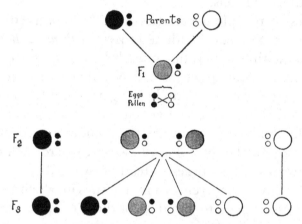

Fig. 90. Diagram of inheritance of red and white flowers in *Mirabilis jalapa* (the Four o'Clock),
the F_1 having pink flowers and the F_2 having red (black circles), pink (grey circles) and white
flowers (white circles) in the ratio 1 : 2 : 1, the pink flowers representing the heterozygous forms,
red being incompletely dominant in this case. (After Morgan.)

strengths of this mixture. On breeding from the pink-flowered cross-
breds, however, it was found that there was no mixture of the genes in
the germ-cells but only of their products in the somatic cells. Thus in the
second generation there appeared on the average out of every four
plants, one red, two pink and one white. Here we get the same ratio as
in the green and yellow peas (p. 3), 25 per cent. pure breeding
dominant red individuals, 25 per cent. pure breeding recessive whites
and 50 per cent. impure breeding individuals carrying red and white,
which in this case gives pink. Thus we see that the genes have retained
their individuality just as surely as in the peas though from the original

cross we might have doubted this. The only difference lies in the balance of the potencies of the genes concerned. In the author's original experiments with fowls many incomplete dominants were found. Out of 1254 dominant individuals observed 466 were complete dominants while 788 were incomplete, giving an approximate 63 per cent. of incomplete dominants in the F_1 generation (Hurst, 1905). In view of Fisher's (1930) classical contribution with calculations of the effect of dominance, in evolution by natural selection, the phenomenon of incomplete dominance, formerly regarded as of little or no importance, is now seen to be of vital importance in evolution. The total "recessiveness" (disappearance) of at least one half of the diploid specific characters in the polyploid species of *Rosa* (Hurst, 1925, 1928) and their reappearance in hybrids further complicates the study of the Fisher effect in evolution.

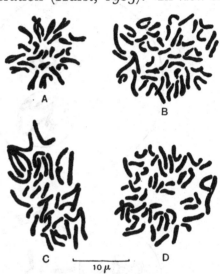

Fig. 91. Diploid and polyploids in *Iris*. A, *Iris variegata* with 24 somatic chromosomes (diploid); B, *I. cypriana* with 48 (tetraploid); C, *Iris* var. "Ballerine" with 36 (triploid); D, *Iris* var. "Ambassadeur" with 48 (tetraploid). Many odd numbers also occur as well as the regular polyploid series (see fig. 49). (After Simonet.)

In the higher animals polyploid varieties appear to be relatively rare in the species so far examined. This is probably largely due to the fact that self-fertilisation is not possible except in some of the lower forms. To produce a race of polyploid animals one individual is not sufficient (except in cases of parthenogenesis) and it is not often that more than one polyploid individual will arise at one time in one place. In a plant the one individual, bearing both egg- and pollen-cells, may self-fertilise and a constant race produced, but in an animal it will have to fertilise or be fertilised by a normal diploid individual, except in remote circumstances, and this will produce irregular numbers of chromosomes and consequently sterility or abnormal offspring which will be sterile also. The best-known case of polyploid varieties in animals is in the Nematode Thread Worm (*Ascaris megalocephala* (fig. 92)) in which the diploid form *univalens* has one gametic and two somatic chromosomes and the tetraploid variety *bivalens* has two gametic and four somatic chromosomes. A cross between these two produces the

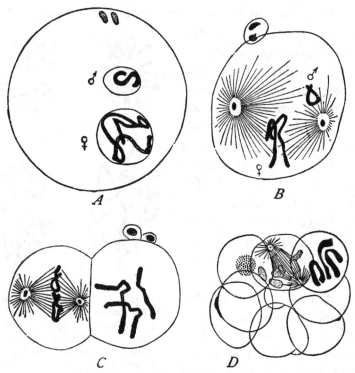

Fig. 92. *A, B, C,* male (♂) sperm from the diploid *Ascaris megalocephala* var. *univalens* with one chromosome, fertilising the egg of the tetraploid var. *bivalens* (♀) with two chromosomes, thus forming a triploid individual with three chromosomes, var. *trivalens*. *D,* a later stage of the embryo: to the right, the primordial germ-cell with the three chromosomes intact; to the left, body-cells with the chromosomes divided up into a large number of small elements. (After Wilson.)

triploid variety *trivalens* with three somatic chromosomes. An octoploid embryo was also observed among progeny of the tetraploid (fig. 93). These forms are remarkable for the chromosomes breaking up in the body-cells but remaining entire in the germ-cells. Right back in the earliest divisions of the individual the cell from which the future germ-cells are to arise is set apart, and in it and all the cells which spring from it the chromosomes remain intact and unchanged. In all the other cells which go to make up the body the chromosomes become sub-divided and fragmented into small elements, and at times lose pieces off their ends in the cytoplasm. This early differentiation of the germ-cells

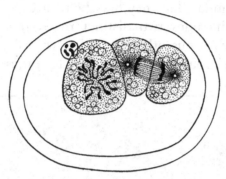

Fig. 93. Octoploid embryo, showing 8 chromosomes of *Ascaris megalocephala* arising by fusion of previously separate eggs of the tetraploid *bivalens*. (After Wilson.)

and the constancy of their chromosomes in *Ascaris* has been one of the chief grounds for belief in the integrity of the germ-cells as compared with the body-cells. The fact that in many animals they are laid down from the beginning and have no real connection with the body except

a   *b*

Fig. 94. Giant germ-cells in Man carrying twice the normal number (48), due to the failure of the chromosomes to make two cells at a division. *a*, germ mother cell with the somatic complement of 96 chromosomes, these coming together at the reduction division, *b*, to form 48 pairs. The occasional functioning of such giant sperms with normal eggs may be the explanation of the rare occurrence of inter-sexes in Man, since in *Drosophila* the triploid condition with two *X* chromosomes gives inter-sexes. (After Painter.)

to take in food supplies and get rid of waste products is most important, and in these animals the actual cytological evidence of this is interesting, since the two types of cell—somatic and gametic—are easily distinguished throughout by their different chromosome contents. A few other cases of this and similar behaviour have been observed in other organisms.

Another well-known case of a polyploid variety in animals is the Brine Shrimp (*Artemia salina*), a cosmopolitan type in which the characters are so similar that they may be regarded as one species. This species, however, includes two distinct races, one from Capodistria (near Trieste) and various other localities, with 84 chromosomes, the other from Cagliari (Sardinia) and other localities, with 42 chromo-

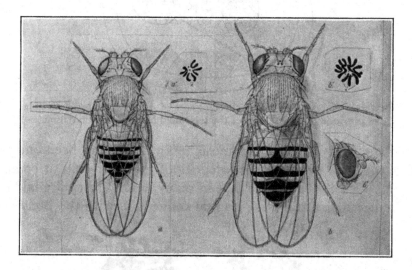

Fig. 95. Normal diploid (*a*) female of *Drosophila melanogaster* and the triploid (*b*), together with their respective chromosome complexes. The diploid has two sets of chromosomes and the triploid three sets. Note increase in size of the triploid variety. (After Morgan.)

somes. The only difference in the two races, so far as external characters are concerned, is the larger size of the nuclei and cells of the tetraploid race and also the larger size of the body.

In the great number of cultures of the Fruit Fly (*Drosophila*), triploid varieties have arisen with 12 chromosomes (fig. 95) and tetraploid varieties with 16 chromosomes. These have been of great value in showing that sex is not due so much to the presence of the sex chromosomes themselves (see p. 35) as in the genic balance of the sex chromosomes with the other chromosomes of the set (autosomes). In triploids, when three X chromosomes are present we get females, because this is the same condition as in the diploid where we have two X chromosomes balanced against two of each kind of autosome, there being three of each kind of autosome in the triploid. If we have two X chromosomes in the triploid against three each of the autosomes the fly is an inter-sex, that is to say, partly male and partly female, while if only one X chromosome is present against three of the others the fly is a super-male. Some cases have arisen in which an extra X chromosome has arisen in the normal diploid by non-disjunction, giving three X chromosomes to two of each of the others. This produced a super-female, but these do not usually survive and if they do they have abnormal ovaries and are

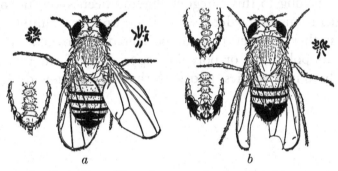

Fig. 96. a, super-female in *Drosophila melanogaster*, showing the chromosome complex with three X chromosomes and two sets of autosomes; b, a super-male with three sets of autosomes and an X and a Y chromosome. (After Bridges.)

sterile. Thus we see that sex is largely determined by the reaction of the genes of the sex chromosomes with the genes of the autosomes and is not due, as was at first thought, simply to the sex chromosomes themselves. Goldschmidt has also demonstrated this in Moths (*Lymantria*). Attempts are being made now to localise the sex differentiator or differentiators in stocks of *Drosophila* by means of different duplications, deletions and translocations within the X chromosome, crosses being made between the different types.

In the above cases of polyploidy the polyploid types are for the most part polyploid varieties of a single species, though since there is a certain barrier of sterility between them and their parents through the changed chromosome complement, the fertile tetraploids, hexaploids and octoploids may be regarded as incipient species with new complements of chromosomes and with greater potentialities for variation by mutation and transmutation than is possessed by the diploids. For example, if translocations occur between the chromosomes of the different sets their identity will be gradually obliterated and a new type will arise with a relatively high chromosome number, and many of the genera to-day which show these higher basic numbers may well have originated in this manner.

These polyploid varieties often represent in plants giants of the diploid species, which are exceedingly useful in many cases as the source of bigger and better varieties for our gardens, and many of our best cultivated plants belong to this category, having been saved unconsciously by past generations of gardeners for their superior qualities. There is, however, another distinct type of polyploidy, known as differential polyploids, allopolyploids or polyploid species, the formation of which leads directly to new species, and with that we will now deal.

CHAPTER IX

POLYPLOID SPECIES

A DIPLOID species has two sets of chromosomes and genes, one derived from each parent. A polyploid species, in common with a polyploid variety, has multiple sets of chromosomes and genes, tetraploids with four sets, pentaploids with five, hexaploids with six, octoploids with eight sets, and so on. Polyploid species, however, differ fundamentally from polyploid varieties, inasmuch as their polyploid sets of chromosomes and genes are those of more than one diploid species while polyploid varieties have the chromosomes and genes of a single diploid species. In other words, polyploid varieties carry the same specific genes in a multiplex state with multivalent chromosomes, while polyploid species carry the genes of more than one species in a duplex state with bivalent chromosomes. If A and B represent the differential sets of chromosomes and genes of two diploid species AA and BB, then $AAAA$ is a tetraploid variety of the diploid species AA, while $AABB$ is a tetraploid species combining the chromosomes and genes of the two diploid species AA and BB. The most widely investigated case of polyploid species is that of the genus *Rosa* (Tackholm, 1922; Hurst, 1928). More than one thousand species and varieties of this genus have been studied cytologically, taxonomically and genetically by nine workers in Europe and America, and all the known species of this polymorphic genus have been investigated.

Roses are very widespread throughout the northern hemisphere, reaching nearly to the equator and extending beyond the arctic circle. Naturally with such a wide distribution there are many different species ranging from the giant diploid Rose (*Rosa gigantea* Coll.), which climbs the trees of the Burmese forests and which, with its long orange-cream flower buds and huge ivory coloured flowers, is one of the glories of the plant world, to the small and insignificant polyploid species, only a few inches high, which grow on the Canadian prairies and in the tundra and arctic regions of Europe, Asia and America.

When the chromosomes of all the different species were examined it was found that though the number was usually constant for each wild species there were several different numbers for the different species of the genus. Species were found with 14, 28, 35, 42 and 56 chromosomes, and hybrids with 21. It will be observed that these numbers are all multiples of 7, so that the genus consists of species which are diploid and polyploid, since all bear chromosome numbers which are multiples of the basic gametic number of the genus. Thus there are diploids with 14 chromosomes, triploids with 21, tetraploids with 28, pentaploids with 35, hexaploids with 42 and octoploids with 56 chromosomes.

The diploid and equal polyploid species with balanced bivalent chromosomes form regular germ-cells and pollen grains with a normal meiosis and gametogenesis which does not differ materially from that of other species of plants and animals.

The unequal polyploid species with unbalanced (bivalent and univalent) chromosomes, consisting of 7 bivalents and either 7, 14, 21 or 28 univalents, have an irregular meiosis followed by a regular but unequal gametogenesis in which the female gametes have twice, three times or four times the number of chromosomes found in the functional male gametes. The pollen grains of these species are polysporic and only those with 7 chromosomes are functional. This remarkable process of gamete formation so far appears to be unique in plants and animals.

The triploids with 7 bivalents and 7 univalents are all hybrids or cultivated varieties which are sterile and apparently cannot persist as species in a wild state.

To the pentaploid species belong the wild Dog Roses and briars of our hedges (*R. canina* L.). These, having unbalanced sets of chromosomes, do not form regular pairs but there are 7 paired bivalents and 21 single univalents in their early germ-cells, male and female. In the pollen mother cells the 7 pairs divide first and irregular pollen grains are formed in which some have only the 7 chromosomes while others contain the 7 with a variable number of the univalents and others again carry only univalents. As a rule only those with the 7 from the bivalents behave normally and produce functional pollen, the end result being that the 21 univalents are discarded and take no part in the male heritage. In the embryo-sac mother cells a remarkable phenomenon appears. All the 21 univalents go to the top of the cell and wait there for the reduced 7 of the bivalents. These together form a cell with 28

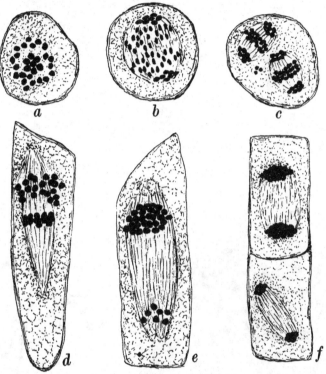

Fig. 97. A comparison of the formation of male and female gametes in the unequal pentaploid species of *Rosa. a, b, c,* pollen mother cells with 7 paired chromosomes which work independently of the 21 single chromosomes, giving male gametes with 7 chromosomes only; *d, e, f,* corresponding divisions in embryo-sac mother cells with regular but unequal reduction of the chromosomes, giving embryo-sacs with 28 chromosomes. *a,* 1st pollen mother cell division metaphase in *R. glaucophylla* Winch, var. *Seringei* (Christ), 7 paired chromosomes in centre, 21 single chromosomes around; *b,* 1st pollen mother cell in *R. uriensis* Lag. et Pug., 7 paired chromosomes reduced, 7 to each pole, the singles splitting; *c,* 2nd pollen mother cell division in *R. orophila* Gren., 7 of the paired chromosomes at each pole to make pollen grains; *d,* 1st division embryo-sac mother cell metaphase in *R. glaucophylla* Winch, 7 paired chromosomes on spindle, 21 singles at upper pole; *e,* 1st embryo-sac mother cell division in *R. elliptica* Tausch, 7 chromosomes below, 28 above (7 from reduced pairs + 21 singles); *f,* 2nd embryo-sac mother cell division in *R. Froebelii* Christ, 7 chromosomes in lower cell, 28 in upper (latter forming the embryo-sac).

chromosomes and this develops to form an embryo-sac containing the egg-cell with 28 chromosomes (fig. 97). Thus we have the egg-cells with

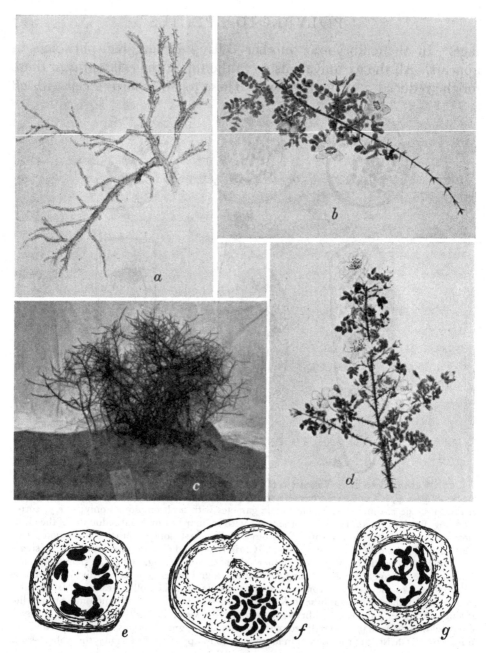

Fig. 98. In the middle, the tetraploid *BBCC* septet species *Rosa spinosissima* L., showing branching (*c*) and flowering (*d*) habits. Above, types of the two diploid species of which it is composed, showing the characters which appear in it: *a*, branching habit of the *CC* diploid species *R. rugosa* Thunb.; *b*, flowering and leaf habit of the *BB* diploid species *R. Willmottiae* Hemsl. Below, chromosomes of the three species: *e*, *R. rugosa* diakinesis with 7 paired chromosomes; *g*, *R. Willmottiae* with 7 paired chromosomes; *f*, pollen grain of *R. spinosissima* with the reduced gametic number (14 chromosomes).

28 and the male nuclei with 7 chromosomes, and when they come to-gether at fertilisation they make a new pentaploid embryo and in-dividual with 35 chromosomes like the parent, in this way completing the cycle of the most remarkable mechanism yet discovered in the formation of gametes.

The unequal tetraploid species with 7 bivalents and 14 univalents and the unequal hexaploid species with 7 bivalents and 28 univalents follow the same mechanism in their gamete formation.

When the taxonomic characters of all the polyploid species were examined and analysed it was found that the wild polyploid species were not, as originally assumed, merely polyploid varieties of the diploid species which had reduplicated their chromosomes, but that they were composed of the characters and chromosomes of several diploid species combined together (Hurst, 1924, 1925). That is to say, in the equal tetraploid species were found the characters and chromosomes of two distinct diploid species, in the equal hexaploid those of three species and in the octoploids the characters and chromosomes of four distinct diploid species. In fig. 98 we see an illustration of this in the tetraploid species *Rosa spinosissima* L., found on the sand dunes of Western Europe. This rose is tetraploid in its chromosomes (28) and an analysis of its taxonomic characters shows that it is composed of the characters of two distinct diploid species represented in the figure by the two Linnean species *R. rugosa* Thunb. and *R. Willmottiae* Hemsl., each of which has 14 chromosomes. The figure shows that the tetraploid species *R. spino-sissima* combines the tortuous, excessively prickly branches of *rugosa* with the singly set flowers and small leaflets of *Willmottiae*, and so on with the other specific characters. Genetical experiments at Cambridge confirm the taxonomic analyses in so far as hybrids between similar Linnean species to the two above present the salient taxonomic characters of *R. spinosissima*.

A comprehensive cytological and taxonomic investigation of all the Linnean species of *Rosa* L., confirmed by many genetical experiments, shows that in *Rosa* L. proper, there are five basic diploid genetical species, each with a differential gametic septet of chromosomes and genes (genome) distinguished as *A, B, C, D* and *E*, thus constituting five

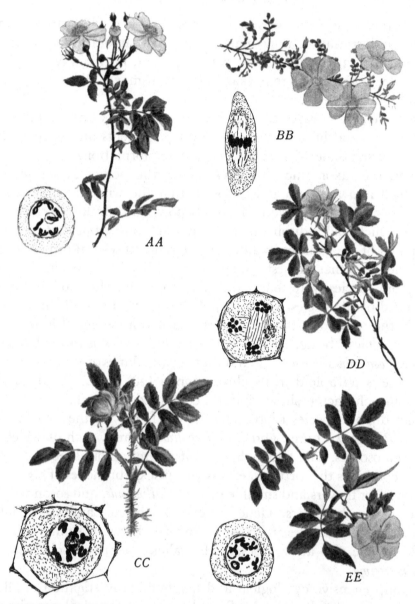

Fig. 99. Representative Linnean species (genetical sub-species) of the five basic diploid genetical species of *Rosa* L. together with their chromosomes in different stages. *AA* septet species *R. arvensis* Huds. with 7 bivalent chromosomes in pollen mother cell; *BB* septet species *R. Webbiana* Wall. with 7 bivalents in embryo-sac mother cell; *CC* septet species *R. coruscans* Waitz. with 7 bivalents in pollen mother cell; *DD* septet species *R. Fendleri* Crép. end of pollen mother cell divisions, showing the reduced number of chromosomes, 7 at each pole; *EE* septet species *R. macrophylla Doncasteri* with 7 bivalents in pollen mother cell.

diploid septet species with somatic chromosomes and genes *AA*, *BB*, *CC*, *DD* and *EE*, respectively (fig. 99). Each of the five genetical species includes a number of closely related Linnean species which by reason of their complete fertility when inter-crossed, and their regional geographical distribution, have been classed as genetical sub-species (Hurst, 1928).

All the wild polyploid species of *Rosa* L. are composed of various combinations of these five diploid septet species. On this basis 26 polyploid septet species with balanced bivalent chromosomes are possible, viz. 10 tetraploids, 10 hexaploids, 5 octoploids and 1 decaploid, and of these 21 have so far been identified in nature, viz. 10 tetraploids, 8 hexaploids and 3 octoploids. With a few exceptions each of these polyploid genetical species corresponds with the Linnean species and all have balanced bivalent chromosomes with regular and normal meioses and gametogeneses.

The tetraploid species *R. spinosissima* L. shown in fig. 98 carries the septets *BBCC*, while the diploid Linnean species *R. Willmottiae* Hemsl. carries *BB* and the diploid *R. rugosa* Thunb. carries *CC* septets, the combination of the chromosomes and genes of the two diploids *BB* and *CC* making the tetraploid species *BBCC*.

The interactions of the differential septet genes in the polyploid species prove to be an interesting study. Certain septet characters usually predominate over the corresponding characters of the other septets, e.g. the exserted styles of the *A* septet; the singly set flowers of the *B* septet; the glandular acicles of the *C* septet; the short straight woolly styles of the *D* septet; and the rimmed pendulous fruits of the *E* septet.

Fig. 100 illustrates the predominance of the singly set flowers in tetraploid, hexaploid and octoploid species carrying the septets *BB*, notwithstanding that in the last two species the *B* septets are in a definite minority. Fig. 101 shows the predominance of the rimmed pendulous fruits in tetraploid, hexaploid and octoploid species carrying the septets *EE*.

With regard to the unequal polyploid species with unbalanced bivalent and univalent chromosomes: omitting those with more than one

Fig. 100. Illustrating the predominance of the *B* septet character, singly-set flowers, in several polyploid Linnean species which carry the *B* septet of chromosomes. *a, Rosa Hugonis* Hemsl. *BB* diploid species with 7 bivalent chromosomes in pollen mother cell; *b, R. ochroleuca* Swartz *BBDD* tetraploid species with pollen grain carrying 14 chromosomes; *c, R. spinosissima* L. var. *duplex* Hort. *BBCC* tetraploid species with 28 somatic chromosomes in body-cell; *d, R. Moyesii* Hemsl. et Wils. *AABBEE* hexaploid species with 21 bivalent chromosomes in pollen mother cell; *e, × R. hibernica* Templ. *ABBCDD* hexaploid natural hybrid with 14 bivalent and 14 univalent chromosomes in pollen mother cell; *f, R. altaica* Willd. *BBDD* tetraploid species with 14 gametic chromosomes in pollen mother cell; *g, R. acicularis* Lindl. p.p. *BBCCDDEE* octoploid species with 28 gametic chromosomes in pollen mother cell.

septet of bivalents which are natural hybrids and omitting triploids which are not found in a wild state as established species, 55 polyploid septet species are possible, viz. 30 tetraploids, 20 pentaploids and 5 hexaploids; of these 20 have so far been identified in nature, viz. 4 tetraploids, 13 pentaploids and 3 hexaploids. With a few exceptions these also correspond with Linnean species and all have 7 bivalent chromosomes together with 14, 21, or 28 univalents, giving rise to irregular meioses followed by regular but unequal gametogeneses. The cytological evidence supports the hypothesis of Tackholm (1922) that these unequal polyploid species with unbalanced chromosomes are ancient F_1 hybrids and the taxonomic and the genetical evidence (Hurst, 1928) fully confirm this hypothesis. According to the geological evidence these unequal polyploid species arose about the middle of the Pleistocene Period after the great Mindel glaciation, and it is a remarkable fact that the present distribution of these species in Europe and Western Asia corresponds rather closely with the area influenced by the advance and retreat of the Mindel ice sheet. The advance of the ice would naturally bring down the surviving arctic polyploid species to mingle with the southern diploid species, thus producing the unequal polyploid F_1 hybrids which have since that time reproduced themselves true to type, mainly by apomictical reproduction. It is interesting to note that the lowland and southern unequal polyploid species *R. canina* L. and *R. micrantha* Sm. carry the bivalent septets *AA* which are also carried by the lowland and southern diploid species *R. arvensis* Huds., while, on the other hand, the alpine and northern unequal polyploid species *R. caesia* Sm. (*R. coriifolia* Fries) and *R. glaucophylla* Winch (*R. glauca* Vill.) carry the bivalent septets *DD* which are also carried by the alpine and northern diploid species *R. cinnamomea* L. 1759. How the equal polyploid species with balanced bivalent chromosomes arose is an interesting problem which may ultimately have several solutions. At present the most feasible and acceptable hypothesis is that the original diploid species of the Miocene and Pliocene were hybridised by insects and that the chromosomes of the hybrids were duplicated, thus producing fertile polyploid species. Genetical experiments confirm this hypothesis so far as the characters of the diploid and polyploid species are concerned, but as yet duplication of the chromosomes in hybrid roses has not been observed under experimental conditions. This may, however, be due to

Fig. 101. Showing the predominance of the *E* septet character, type of fruit, in *Rosa* polyploids. 1, 2, fruits of the *EE* diploid species *R. macrophylla* Lindl.; 3, *R. Fargesii* Hort. (tetraploid *AAEE*); 4, *R. Davidii* Crép. (another Linnean tetraploid *AAEE*); 5, *R. Moyesii* Hemsl. and Wils. (hexaploid *AABBEE*); 6, *R. acicularis* Lindl. p.p. (octoploid *BBCCDDEE*).

the recognised technical difficulties peculiar to the sexual reproduction of woody shrubs, since in genetical experiments with herbaceous genera at least seventeen cases are definitely known where two species of plants have been hybridised and a subsequent duplication of the chromosomes has produced new fertile polyploid individuals which in every respect satisfy the taxonomical requirements of a new species. These individuals constitute an entirely new species, inasmuch as in combining the characters of two distinct species a new end result is achieved which is distinct from either of the parent species. Further, the increased number of chromosomes isolates them sexually from their parents, since though quite fertile with one another they produce sterile triploids back-crossed with either parent. In this way new species and in a few cases new genera have arisen under experimental control, and there can be no doubt that the seventeen recorded cases provide an experimental demonstration of the evolution of new species.

The original case of this kind of formation of a new species was *Primula kewensis* (fig. 102). At Kew two diploid species *P. verticillata* and *P. floribunda* crossed spontaneously and produced a completely sterile hybrid × *P. kewensis*. This was cultivated vegetatively by division for some years when it suddenly produced a fertile shoot which flowered and bore normal seeds. A cytological examination of this showed it to be tetraploid in its chromosomes, the original species being diploid with 18 somatic chromosomes as was the sterile hybrid, while the new fertile species had 36 chromosomes. Thus arose a new and fertile species *P. kewensis* with double the number of chromosomes and the combined characters of both the parent species (Pellew and Newton, 1929).

Another case arose in the Tobacco Plant (*Nicotiana*) in the genetical cultures of Goodspeed and Clausen (1925), although in this case the doubling of the number of chromosomes took place in the germ-cells and not in the body-cells. Crossing *N. glutinosa* (24 chromosomes) with *N. tabacum* (48 chromosomes), sterile hybrids arose with 36 chromosomes. From one of these was raised an individual with twice the number of chromosomes (72). This plant, being fertile, was bred from and has maintained itself absolutely true for five generations (fig. 103).

Since it has all the attributes of a true species it has been given a specific name—*N. digluta*. Clausen, working with *N. tabacum*, the cultivated Tobacco Plant, thinks that this species originated by a similar

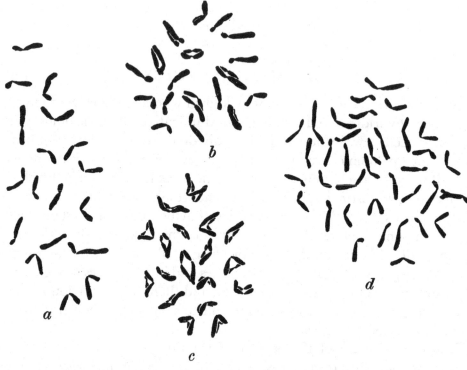

Fig. 102. Chromosomes of *Primula*. *a*, *P. floribunda* (18 somatic chromosomes); *b*, *P. verticillata* (18 chromosomes); *c*, ×*P. kewensis*, diploid hybrid between *a* and *b* (18 chromosomes); *d*, *P. kewensis*, tetraploid species arising from *c* (36 chromosomes); *e*, plant of the diploid hybrid, showing (right) the crown with broad leaves from which a fertile inflorescence arose having the tetraploid number of chromosomes, illustrating the case of a new species arising as a bud transmutation. (After Newton and Pellew.)

cross between two diploid species *N. sylvestris* and *N. tomentosa* followed by duplication, since each of these has only 24 chromosomes while *tabacum* has 48. Comparison of the characters shows that duplication has been followed by certain alterations in the gene contents of the chromosomes, which have not yet, however, destroyed their cytological

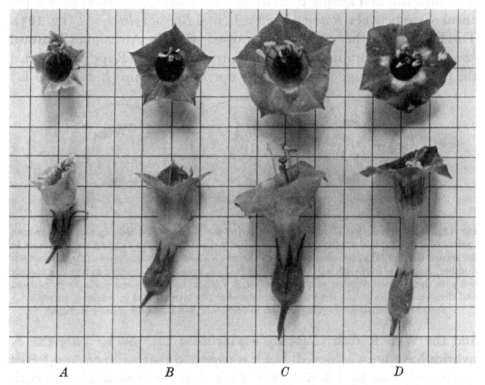

A   B   C   D

Fig. 103. *A, Nicotiana glutinosa* (12 pairs of chromosomes) and, *D, N. tabacum* (24 pairs) which, crossed together, gave the F_1 (*B*) with 36 chromosomes, which, duplicating, gave the constant species *N. digluta* (*C*) with 36 pairs of chromosomes. *N. digluta* appears as a merely enlarged form of the F_1 hybrid, except that it is abundantly fertile. (After Clausen.)

affinity with the original diploid species, since in crosses with either, 12 paired chromosomes and 12 singles occur. In a state of nature the new species, by reason of its complete fertility, would no doubt prevail against any semi-sterile back-crosses with the parent species if all three grew together in mutual competition. In the above cases when the gametes were formed some arose containing the entire sets of chromosomes, which either had not reduced in the normal way, or had fused in the second division (fig. 104); these, coming together, formed fertile plants with twice the number of chromosomes of the hybrid parent.

An even more striking case is the experimental creation of a new genus by crossing the two genera *Brassica* (Cabbage) and *Raphanus* (Radish), a fertile tetraploid arising from the union. The original cross by Karpechenko was between *Raphanus sativus* L. and *Brassica oleracea* L. (fig. 105). This has been further successfully crossed with the Abyssinian Mustard (*Brassica carinata* Braun), the Swede Turnip and Colza Rape (*B. napus* L.) the Chinese Cabbage (*B. pekinensis* Rupr.), the Turnip (*B. campestris* L.)

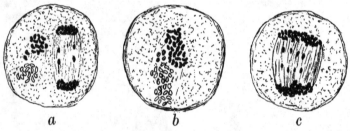

a   *b*   *c*

Fig. 104. The production of gametes with twice the number of chromosomes in a garden variety of the tetraploid species *Rosa pendulina* L. by the fusion of the reduced chromosomes in the second division, thus forming two pollen grains each with 28 chromosomes (the full somatic number) instead of four with half that number (14). *a*, normal 2nd division in a pollen mother cell with the reduced number of chromosomes; *b*, *c*, polar and side views respectively of fused 2nd division spindles giving the full somatic number of chromosomes in resulting gametes. A few chromosomes are lagging behind in each case, but these are usually included.

and the Wild Radish (*Raphanus raphanistrum* L.). So far only the F_1 hybrids have been achieved, but it is expected that new constant fertile species will arise by a duplication of the chromosomes, which will each contain the characters of the three species involved in the hybridisation and thus provide experimental proof of the origin of new hexaploid species in this manner. Russian geneticists have also experimentally raised a second new genus *Triticale* by crossing Wheat (*Triticum vulgare*) with Rye (*Secale cereale*). The wheat parent was a hexaploid species with 42 chromosomes and the rye a diploid species with 14. The sterile hybrid was a tetraploid with 28 chromosomes, which, duplicating, gave the new fertile octoploid genus Wheat-Rye (*Triticale*) with 56 chromosomes. In this way one may get fertile progeny from the most unlike

Fig. 105. Above, regular chromosome divisions (18 pairs) in the gametes of the new tetraploid genus *Raphanobrassica*, arising from the sterile diploid F_1 hybrid *Raphanus* × *Brassica* (Radish × Cabbage). Below, pods of the tetraploid (left) showing complete fertility compared with the sterile diploid hybrid (right). (After Karpechenko.)

crosses provided that the cross can be achieved in the first place, and this creation of new species and new genera, as we have seen, is and has been no doubt a potent factor in evolutionary progress.

One of the most interesting cases of the creation of a new species in this manner is that of the Pink Chestnut (*Aesculus carnea* Willd.). Originally a cross between two species of distinct sections of the genus (*A. Hip-*

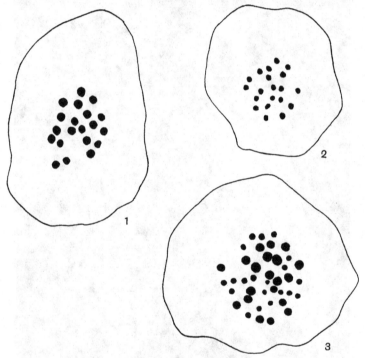

Fig. 106. New species formation in the Horse-chestnut. 1, *Aesculus Pavia* with 20 pairs of chromosomes; 2, *A. Hippocastanum* with 20 pairs of chromosomes; 3, the hybrid between the two with 40 pairs, 20 large from *Pavia* and 20 small from *Hippocastanum* duplicated, giving a fertile new species, *A. carnea* (the Pink Chestnut.) (After Skovsted.)

pocastanum, the Horse Chestnut, from Eurasia and *A. Pavia* from North America) it aroused considerable interest by its fertility and its breeding true from seed. 218 seedlings raised by the author showed slight variations in the amount of yellow at the base of the flowers and other minor differences. Otherwise they bred true and the plant is often propagated by this means. The unusual fact of its pure breeding has been recently explained by an investigation of the chromosomes. Skovsted (1929) finds that while the parents have each 20 pairs of chromosomes the hybrid has 40 pairs, in short the chromosomes have doubled and thus formed a constant true-breeding species combining the characters of two sections of the genus. Of exceptional interest is the fact that *A. Pavia* has 20 large chromosomes and *A. Hippocastanum* 20 small ones

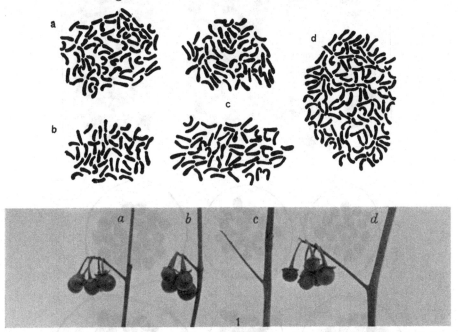

Fig. 107. New tetraploid *Solanum* species, chromosomes and fruits. *a, S. nigrum*; *b, S. luteum*; *c, S. luteum × nigrum* diploid hybrid (sterile, no fruits) and, *d*, the new fertile tetraploid species *S. luteo-nigrum* arising from it by a duplication of the chromosomes. (After Jörgensen.)

in the germ-cells while the hybrid has 20 large and 20 small, showing definitely the existence of the chromosome complex of each parent side by side (fig. 106). The slight variation in flower colour is analogous to some slight variations which also occur in *Primula kewensis*. This may be due to a change of gene constitution in one chromosome after the duplication, or it may be due to two chromosomes of the parents being like enough to join up and in the duplicated form to give a tetrasome with random assortment of the four constituents, thus giving segregation in those genes carried by these particular chromosomes.

Several other genera which have been examined have shown a similar series of polyploid "species" to those in *Rosa*, though whether these are polyploid species or polyploid varieties has not yet been

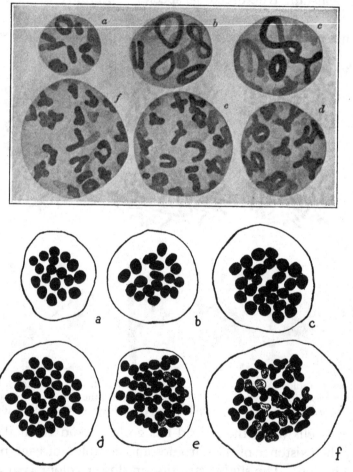

Fig. 108. Polyploid species and varieties of *Chrysanthemum*, with 9 as the basic gametic number of chromosomes. *a*, *b*, tetraploids with 18 gametic chromosomes (36 somatic); *c*, hexaploid with 27 (54 somatic); *d*, octoploid with 36 (72 somatic); *e*, *f*, decaploids with 45 (90 somatic). (After Tahara.)

critically determined. Japanese botanists have found polyploid series in the genus *Chrysanthemum*, but in this case the chromosomes are in multiples of 9. Species with 18, 27, 36, 54, 72 and 90 have been found—

that is, diploids, triploids, tetraploids, hexaploids, octoploids and decaploids (fig. 108). It is an interesting fact that in *Rosa* the species in the
extreme south towards the equator are all diploids, while the polyploids
have a more northerly distribution, the octoploid species *R. acicularis*
Lindl., which is the highest polyploid species yet found in this genus,
being, so far as critically known, usually arctic and circumpolar in its
distribution, though from living material collected for me by Prof.
Cockerell it apparently extends south to Lake Baikal in Siberia, where
conditions are sub-arctic. Similarly, in *Chrysanthemum* the decaploid
with 90 chromosomes, the highest found, appears also to be an arctic
species (*C. arctica*). Work on several different species of this genus by
Shimotomai (1931) points to the fact that there are polyploid species
and polyploid varieties in this genus as in *Rosa*, but much more remains
to be done to show all the relationships between them.

In the genus *Senecio* (Groundsel) Afzelius (1924) found in forty species
a multiple series of somatic chromosomes, 10, 20, 40, 50, 60, 180, and
in twenty species of ten genera in the same tribe (Senecioneae) he found
a similar series 10, 20, 40, 50, 60, so that all these kindred genera have
apparently followed the same mechanism of evolution with 5 as the
basic number of the tribe. It is remarkable how constant the chromosome numbers remain in this tribe—always in multiples of 5—and the
correlation between the chromosomes and the characters is clearly close.

In another tribe of the Compositae family, Tahara and Shimotomai
found the basal number to be 9 for *Aster* and certain related genera,
giving diploids with 18 somatic chromosomes, tetraploids with 36 and
hexaploids with 54.

Other genera of the family Rosaceae besides *Rosa* have multiple
numbers. *Rubus* (Raspberries and Blackberries) has precisely the same
numbers as *Rosa*, the basic number being 7. In this genus also there are
diploids with 14, triploids with 21, tetraploids with 28, pentaploids with
35, hexaploids with 42 and octoploids with 56 chromosomes. In *Poten-
tilla* (Cinquefoils) diploids with 14 and tetraploids with 28 chromosomes
have been found. *Fragaria* (Strawberries) also has septets of chromosomes, in this case diploids with 14, hexaploids with 42 and octoploids
with 56 having been found. In *Prunus* (Plums and Cherries), however,

the basic number changes to 8 (fig. 109), while in *Alchemilla* (Lady's Mantle) the lowest number yet found is 16 (gametic), with "tetraploid" species with 64 chromosomes. If 8 is also the basic number in this genus the diploids have been washed out or the genus may have originated from an old tetraploid which had come to resemble a diploid species

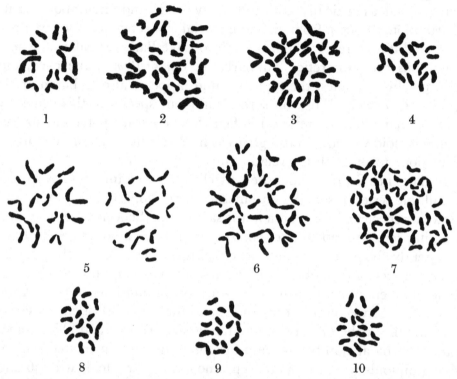

Fig. 109. Diploid, tetraploid and hexaploid species and varieties of *Prunus*. 1, *P. avium* var. (Sweet Cherry) with 16 chromosomes; 2, *P. cerasus* var. (Sour Cherry) with 32; 3, *P. fruticosa reflexa* (Ground Cherry) with 32; 4, *P. mahaleb* (Mahaleb Cherry) with 16; 5, *P. cerasifera* (Cherry Plum) with 16; 6, *P. spinosa* (Blackthorn or Sloe) with 32; 7, *P. insititia* ("King of the Damsons") with 48; 8, *P. persica* (Peach) with 16; 9, *P. amygdalus* (Almond) with 16; 10, *P. triflora* (Japanese Plum) with 16. The edible varieties of the Sweet Cherry (*P. avium*) which have been examined are hyperploid, i.e. have the diploid number plus 1, 2, or 3 extra chromosomes giving numbers from 17 to 19. (After Darlington.)

by mutations and segmental interchanges between the chromosomes of the two sets.

The Pomoideae sub-family of the Rosaceae, including *Pyrus* (Apples, Pears, White Beam, Mountain Ash, and Service Tree), *Crataegus* (Hawthorns), *Cotoneaster*, *Pyracantha*, *Cydonia* (Quince), *Mespilus* (Medlar), *Amelanchier* and other less known genera, are remarkable in having 17 as their gametic number of chromosomes. Recent work by Darlington and Moffett on these genera has shown conclusively that this unusual number has arisen from ancestral species with 7 chromosomes. They find in their cytological preparations that of the 34 chromosomes present in the "diploid" (fig. 110) 4 of the chromosomes are represented four times and 3 chromosomes are represented six times, since, when the gametes are being formed, instead of finding 17 paired chromosomes as ex-pected, there are frequently different numbers of multivalent chromosomes in secondary association and in extreme cases there are four groups containing 4 chromosomes in each and three groups containing 6, making only seven groups in all (fig. 111). Thus these Pomoideae are "trebly hexasomic tetraploids", and the number 17 is a secondary basic number, having arisen by the duplica-tion of the original 7 chromosomes and the subsequent reduplication, probably by non-disjunction, of three more pairs of chromosomes. The polyploids arising from it are termed secondary polyploids.

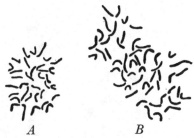

Fig. 110. Somatic chromosomes in Apple varieties (*Pyrus Malus*) in which 17 is the gametic number. *A*, Worcester Pearmain with 34 ("diploid"); *B*, Blen-heim Orange with 51 ("triploid"). All cultivated apples so far examined are either "diploid" or "triploid," but three species (*P. Toringo*, *P. glaucescens* and *P. coronaria*) are "tetraploid". (After Crane and Lawrence.)

This is an especially interest-ing demonstration of the origin of a new number of chromosomes and, judging from the effect upon the characters of the extra chromo-somes in the transmutants that have arisen in cultures with one or more additional chromosomes (p. 82), it is highly probable that this sub-family owes its different fruit structure (pome) and other peculiarities, which distinguish it from the rest of the family, to its extra chromo-somes, showing a new family in process of evolution. Those members of the Rosaceae with 8 or 16 chromosomes may have branched off by non-disjunction from the original 7 chromosome ancestors, which number still persists, as we have seen, in a large part of the family.

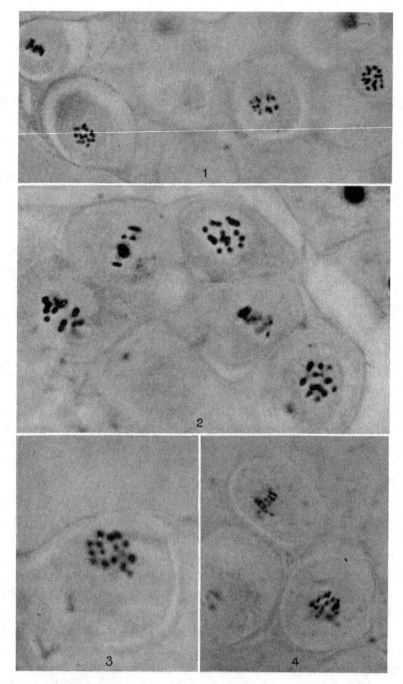

Fig. 111. Microphotographs of pollen mother cells in Apples, showing sexivalent, quadrivalent and bivalent chromosomes associations. 1, 2, Northern Spy; 3, 4, Blenheim Orange. (After Darlington.)

An analogous number occurs in the Salicaceae (Poplars and Willows), where Blackburn and Harrison (1924) found 19 as the basic number, with diploids having 38, tetraploids 76 and hexaploids 114 somatic chromosomes. Since the chromosome complement of these species contains some large, apparently composite, chromosomes, the authors consider that this number is probably due to the fusion of several chromosomes in ancestors which had a more normal number. In support of this two other species were found, one of which had 22 (44 somatic chromosomes) and the other the tetraploid number 44 (88 somatic), the increase of number from 19 to 22 being considered as due to the splitting up of the compound chromosomes.

Species of the Whitlow-Grass, *Draba*, of the family Cruciferae, have 16, 24, 32, 40, 48, 64 and 80 chromosomes, 8 being the basic number (Heilborn, 1928). Only a fraction of the genus has yet been worked out, but one species, *D. magellanica*, spread over four continents, shows the series 48, 64, 80, and this may be a case of duplicated polyploid varieties. On the other hand we have the interesting case of *Rosa acicularis*, which is said by different cytologists to have diploid, tetraploid, hexaploid and octoploid forms. A critical taxonomic diagnosis of the material used shows clearly, however, that the original *R. acicularis* of Lindley is octoploid and includes two distinct genetical species with septets *AACCDDEE* and *BBCCDDEE*, while the American hexaploid "acicularis" are either *R. Bourgeauiana* Crép., which is *BBCCDD*, or *R. Sayi* Schwein., which is *CCDDEE*. The tetraploid "acicularis nipponensis" of Willmott (non Crép.) found in the Kew collection is a sub-species of *R. pendulina* L., which is *DDEE*, while the original diploid *R. acicularis nipponensis* of Crépin is a sub-species of *R. rugosa* Thunb., which is *CC*.

In the genus *Rumex* (Docks and Sorrels) another long polyploid series of chromosome numbers has been found, 10 being apparently the basic number, and there are species with 20, 40, 60, 80, 100, 120 and 200 chromosomes in their body-cells.

Some genera, such as *Viola* (fig. 112) and *Crepis*, appear to conform with both rules, having some of their species which present a series of multiple and polyploid numbers, while others differ by only one or more pairs of chromosomes as found in *Carex* and other genera.

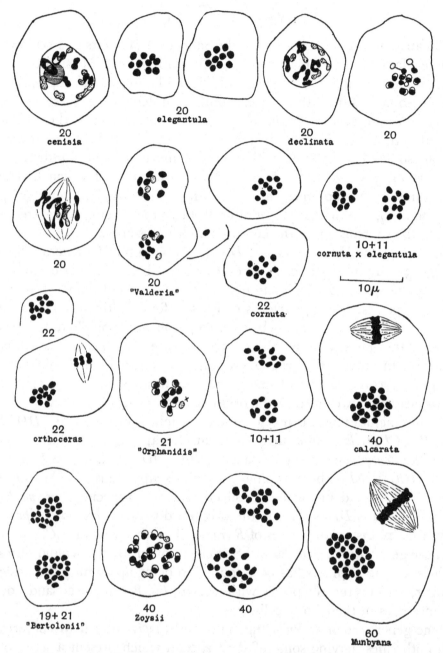

Fig. 112. Chromosomes of the section *Melanium*, group *Calcaratae*, in *Viola*, in which is found a series with 10 as the basic number. Another group of this section falls into a 6 series, while other sections of the genus form 6, 10 and 12 series. The name of the species is below together with the chromosome number. *V. cornuta* has 11 gametic and 22 somatic chromosomes, the extra pair probably arising by non-disjunction. The hybrid with *elegantula* consequently has 10 + 11 chromosomes, *V.* "*Orphanidis*" is the same. (After Clausen.)

An interesting case of polyploid series of chromosome numbers in different species is found in the monocotyledonous cereals, Wheat (*Triticum*), Oats (*Avena*) and Barley (*Hordeum*). The Wheats especially have been systematically investigated in their taxonomy, genetics and cytology (Watkins, 1931). In many respects they resemble the Roses, their basic number of chromosomes is 7 and their sets of chromosomes are septets; their polyploid species also appear to be composed of different combinations of the chromosomes and genes of several diploid

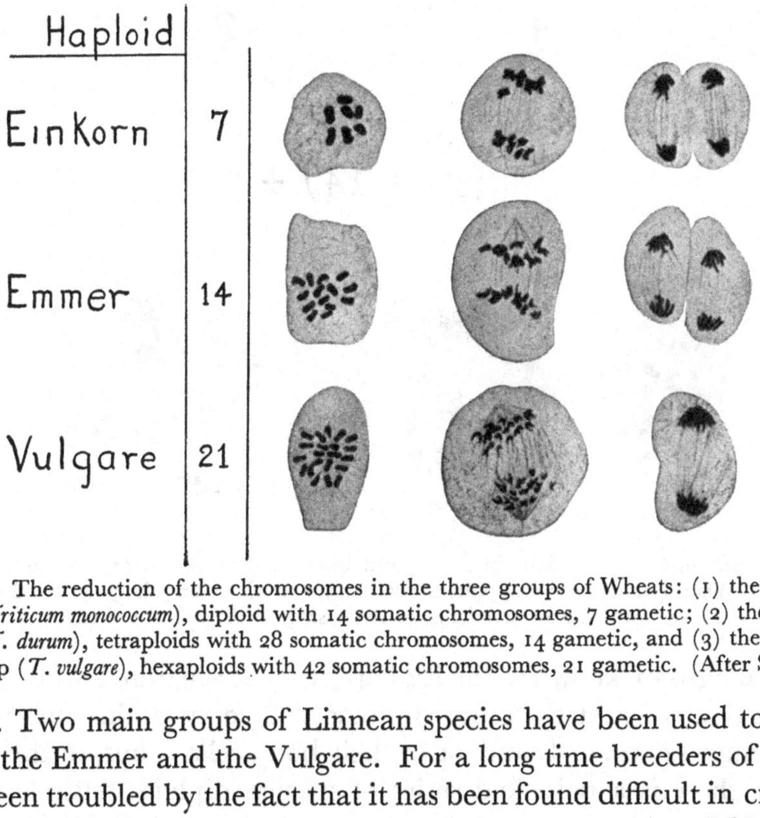

Fig. 113. The reduction of the chromosomes in the three groups of Wheats: (1) the Einkorn group (*Triticum monococcum*), diploid with 14 somatic chromosomes, 7 gametic; (2) the Emmer group (*T. durum*), tetraploids with 28 somatic chromosomes, 14 gametic, and (3) the Vulgare group (*T. vulgare*), hexaploids with 42 somatic chromosomes, 21 gametic. (After Sax.)

species. Two main groups of Linnean species have been used to make bread, the Emmer and the Vulgare. For a long time breeders of wheat have been troubled by the fact that it has been found difficult in crossing the Emmer with the Vulgare wheats to form a new race on Mendelian lines with the good qualities of both races. An examination of the chromosomes explains the difficulty (figs. 113, 114). The Emmer group have 14 chromosomes in their reduced germ-cells and 28 in their body-cells, while the Vulgare group have 21 in their reduced germ-cells and 42 in their body-cells. In crossing varieties of the two groups the unequal number of chromosomes causes a great deal of sterility and in the following generations the univalent chromosomes which cause the sterility tend to disappear in the different divisions, and since certain of the genes representing the good characters required are located in the univalents, they also tend to disappear. The Emmer group of Linnean species are all tetraploids and the Vulgare group are all hexaploids, each group including distinct genetical species as in *Rosa*.

$$14 \qquad\qquad 21$$

$$(14 + 14) + 7$$

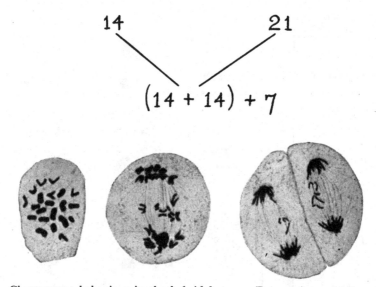

Fig. 114. Chromosome behaviour in the hybrid between Emmer (tetraploid) and Vulgare (hexaploid) Wheats, showing the 14 pairs which behave regularly and the 7 unpaired chromosomes which behave irregularly causing sterility. (After Sax.)

Another group, known as the Einkorn wheats, are from a wild species, little cultivated, and these are diploids with only 7 chromosomes in their reduced germ-cells and 14 in their body-cells. Thus, in the genus *Triticum* there is a polyploid series of species, diploid, tetraploid and hexaploid, while recently a new octoploid genus has been created by Tschermak (1926) by crossing a tetraploid species of *Triticum* with a tetraploid species of *Aegilops*. The sterile tetraploid hybrid produced by this cross duplicated its chromosomes as in the radish-cabbage and the wheat-rye hybrids, thus forming a new fertile octoploid genus known as *Aegilotricum*. More recently Tschermak (1930) has succeeded in raising several new species of *Triticum*, including a hexaploid species *T. turgido-villosum* from a hybrid between a tetraploid and a diploid species, the F_1 being a sterile triploid hybrid which duplicated its chromosomes and thus produced a new fertile hexaploid species.

The above cases demonstrate experimentally the nature of polyploid species and show definitely how new species may arise in nature with relative rapidity through the natural hybridisation of diploid species by insects and other agencies. The fact that these new polyploid species and other transmutations have originated in genetical experiments under critical human control opens up a new chapter in the history of evolution and places the study of evolution on the sound basis of an experimental science.

CHAPTER X

HYBRIDISATION AND CROSSING

ONE of the four great factors in the evolution of species has un-doubtedly been the occurrence of frequent hybridisations and crossings. By natural and experimental crossings in plants and animals it is possible to get numerous recombinations of new and old characters and new mutations and transmutations in related and inter-fertile varieties, while in the hybridisation of Linnean species and genera many chromosome transmutations arise, some of which, being viable and fertile, give rise to entirely new species and genera. As we have seen in the previous chapter, many new polyploid species and in three cases new polyploid genera (fig. 116) have arisen from the hybridisation of unlike diploid and polyploid species which have produced either diploid gametes or duplication of the chromosomes in the somatic cells, and have given rise to new polyploid species with regular chromosome be-haviour and consequent fertility. In the natural crossing of unlike varieties and species, new varieties and hybrids of plants and animals may arise which, becoming sexually isolated from the original parent species by chromosome or gene differences and infertility or by geo-graphical and ecological barriers, may give rise to more and more diverse varieties which, in the course of time, may develop into distinct species.

Many complications occur in hybridisation. In most cases where the individuals belong to different tribes or families it is not possible to effect any union at all, this often being due to incompatibility of the structure of the individuals concerned. In other cases, where the union is possible and fertilisation takes place, there is either no result from the fertilisation or the embryo grows for a short time and then dies. This latter condition is due primarily to the unlike or incomplete chromo-some sets not being able to work together to produce a normal indi-vidual. Different animals and plants have different rates of cell

LATE BLACK "B"

	Flowers			Fruits
A	317	×	Big de Schrecken	142
B	250	×	Kentish Big	109
C	174	×	Ludwigs Big	67
D	{83	×	Wye Morello	30{
	{65	×	May Duke	24}
E	321	×	Turkey Heart	0
F	252	×	Selfed	0

Fig. 115. The Sweet Cherry (*P. Avium*) var. "Late Black", on which various crosses were made, showing the different results. This variety is self-sterile, shown by the branch *F*, which was self-pollinated. Other branches show *E*, cross-incompatibility; *A*, *B*, *C*, cross-compatibility and, *D*, fruits from interspecific pollinations. Below is a list of pollinations (left) and resultant fruits (right). (After Crane and Lawrence.)

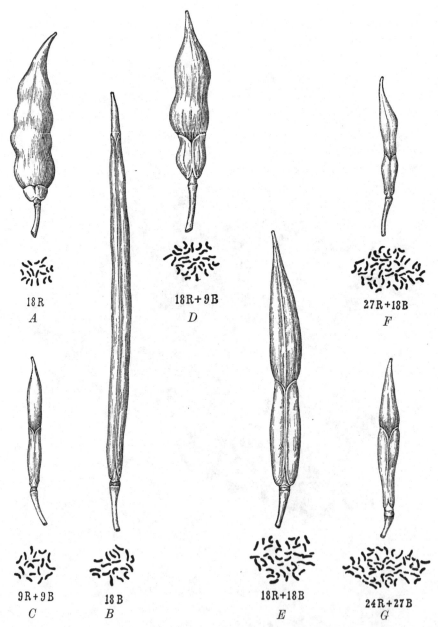

Fig. 116. The creation of a new genus from the hybrid between *Raphanus* (Radish) and *Brassica* (Cabbage). Seed-cases, somatic plates of chromosomes and formula (showing number of chromosomes included of each species) of *A*, *Raphanus*; *B*, *Brassica*; *C*, the sterile diploid hybrid; *E*, the fertile tetraploid species arising by the duplication of the chromosomes of the hybrid; *D*, the triploid arising from a back-cross of *E* with *A*; *F*, a pentaploid arising from a tetraploid gamete of *C* back-crossed with *A*; *G*, a hypohexaploid from the fusion of diploid and hexaploid gametes of *C*. Note effect of more and less chromosomes of radish or cabbage on the form of the seed-case. (After Karpechenko.) (See fig. 105.)

division and it is easy to see that when two sets of chromosomes are brought together having different rates, a time will soon come when one set is working very much behind the other and in consequence cell

Fig. 117. A hypoenneaploid with 78 chromosomes which arose in later generations from *Raphano-Brassica* hybrid by crossing a triploid with a hypohexaploid hybrid (see fig. 116). (After Karpechenko.)

division becomes more and more irregular until it ceases altogether. In other cases the female cytoplasm in which the embryo is developing, either in the animal mother or in the ovule of the plant, has an adverse effect upon the male elements which causes their death. In a few isolated cases of this latter nature the maternal cell is roused to action by the presence of the male nucleus, although the latter does not function, and a haploid or diploid individual is produced by the growth of the mother cell alone.

Frequently the unlike chromosome sets are able to work sufficiently well together to produce an adult individual which is not, however, capable of producing germ-cells, or if germ-cells are produced they are entirely sterile, the sterility being caused as a rule by the inability of the chromosomes to find mates and to carry through regular divisions. As we have seen, in a few cases they perform an ordinary equatorial division and so form gametes with the full somatic complement of chromosomes which may produce fertile polyploid individuals. In several *Rosa* hybrids abnormalities appear in the pollen grains themselves. Each original germ-cell produces four pollen grains, each with half the number of chromosomes. In these hybrids the four grains do not separate properly but come together and fuse, in some cases two, in others three and in others the whole four nuclei joining together. If these function they will produce high polyploid progeny. In a tetraploid hybrid a four-grained fusion would give rise to a decaploid if fertilising a normal egg-cell. In a diploid, a pentaploid would arise and this may be the origin of certain pentaploid tulips (fig. 118), since de Mol found similar abnormalities in the pollen grains of some garden varieties of this genus. Most of the wild pentaploid roses, however, are more complex in their characters than this origin would warrant, being allopentaploids composed of four distinct diploid species.

Thus in hybrids it is possible to get all degrees of sterility and fertility, all states being traceable in most cases to the possibility or impossibility of the chromosome sets and their contained genes to work together. The favourable or unfavourable influence of maternal cytoplasm and endosperm must also be considered. In some plants, for instance, the pollen tubes, which grow out from the pollen grains to carry the male nuclei down the styles to fertilise the egg-cells, are killed as they proceed through the female tissue. Even in some normal cases where the pollen is carrying two different genes half the pollen, which is carrying the one gene, will grow much more rapidly than the other half, into which the

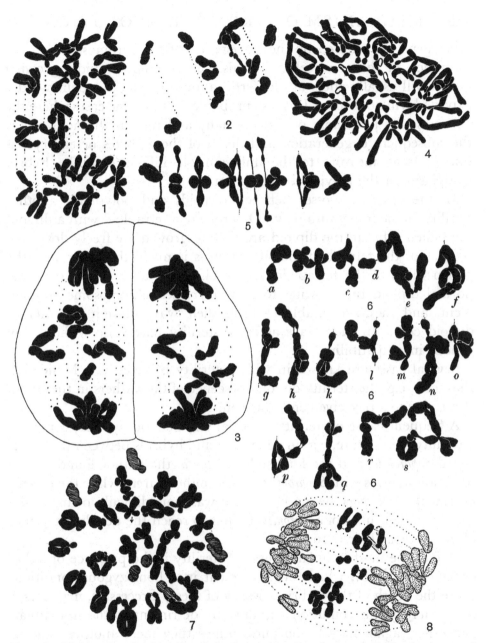

Fig. 118. Divisions in triploid (36 somatic chromosomes) and pentaploid (60 chromosomes) Tulips. 1, gametic division in Keizerkroon, a triploid showing lagging chromosomes; 2, 3, Massenet, another triploid; 4–8, divisions in *Tulipa Clusiana*, a pentaploid; 4, somatic; 6, different types of chromosome conjugation *a, c, e, f, g, h, k, l, o, p*, trivalents; *b, d, m, n, q*, quadrivalents; *s*, a quinquevalent; 7, gametic chromosomes showing 1 quinquevalent, 2 quadrivalents, 3 trivalents, 15 bivalents (black), 8 univalents (hatched). (After Newton and Darlington.)

other gene has entered by the segregation of the chromosomes at the formation of the germ-cells. In this way the more rapidly growing pollen will usually effect all the fertilisations and the Mendelian ratios expected will not occur. It is also probable that in some cases where the chromosomes pair more or less normally and regular divisions occur, the subsequent degeneration and death of the gametes or embryos is due in large measure to the unfavourable influence of the hybrid cytoplasm, or the polyploid endosperm.

In the cases of crosses between diploids and polyploids another sterility factor comes into play. On crossing a tetraploid with a diploid the pollen tubes of the diploid are able to grow quite freely down the stigmas of the tetraploid, but if the cross is made the other way the pollen tubes of the tetraploid are much larger and stouter than the normal diploid tubes owing to their increased chromosome complement, and they are unable to push their way through the smaller diploid cells. In such ways as these sterility may arise without any genetic incompatibility.

Several cases are known in which sterility behaves as a Mendelian character, e.g. the sterility or contabescence of the anthers in the Sweet Pea is due to a recessive gene (Bateson, 1913).

A frequent cause of sterility, especially in generic hybrids, is incompatibility of structure, e.g. in generic orchid hybrids between the short-styled *Sophronitis* and the long-styled *Cattleya*, the author found it easy to obtain hybrids with *Sophronitis* as the mother parent while the reverse cross with *Cattleya* as the mother parent was more difficult, owing to the shorter pollen tubes of *Sophronitis* failing to reach the deeply set ovules of the *Cattleya* (Hurst, 1925).

One of the most frequent causes of sterility is the presence of lethal genes (see p. 178), which, when present in a homozygous condition, cause the death of the embryo. Many of these, if recessive, may persist in a heterozygous condition without any detriment to the individual, and others again may be harmless unless they meet another gene in combination with which a lethal effect is produced. In this latter manner sterility frequently arises between what would otherwise be compatible varieties, two of which may carry two such genes which alone are harmless, but on crossing, the combination of the two causes complete sterility or the death of the zygote. This may also occur in self-fertilisation in some cases.

One of the most remarkable proofs of the influence of cross-breeding in evolution lies in the extensive range of new varieties which have been produced by means of crossings and hybridisations in our domesticated animals and cultivated plants. Many changes here of course are due to mutations (see p. 171) and transmutations (fig. 119), as in all organisms, but a great deal of the variety is due to new recombinations by cross-breeding and the careful selection of the progeny. Some authors do not consider these results to have any significance in evolution, since most of them have been caused by man's intervention and human selection, but on the other hand they give definite proof of the capabilities of organisms to change in all directions, and one must realise that in many cases abnormalities and even monstrosities have been selected as more useful or fanciful for the breeders' purposes while the more normal mutational changes have been discarded as not being of sufficient interest. In recent years much interesting work has been done on cultures of wild species (e.g. Marsden-Jones, 1930),

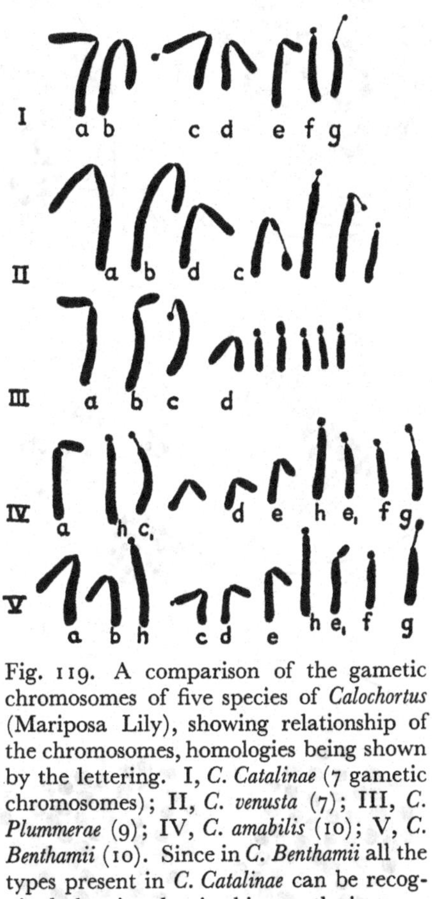

Fig. 119. A comparison of the gametic chromosomes of five species of *Calochortus* (Mariposa Lily), showing relationship of the chromosomes, homologies being shown by the lettering. I, *C. Catalinae* (7 gametic chromosomes); II, *C. venusta* (7); III, *C. Plummerae* (9); IV, *C. amabilis* (10); V, *C. Benthamii* (10). Since in *C. Benthamii* all the types present in *C. Catalinae* can be recognised, showing that in this case the increase of number cannot be due to fragmentation, it is suggested that the extra chromosomes are the result of crossing followed by duplication of certain chromosomes as found by Clausen in his *Viola* experiments. (After Newton.)

and there have been raised by hybridisation a large number of entirely new species which are capable of maintaining themselves permanently

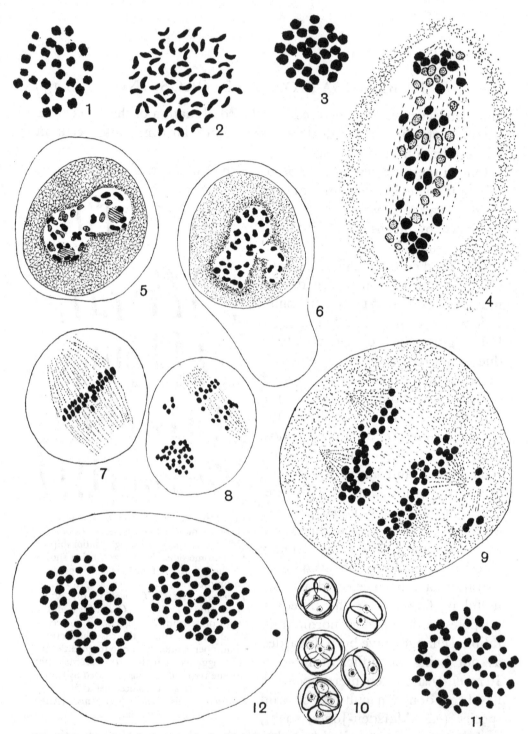

Fig. 120. Creation of a new species in *Digitalis* (Foxglove). 1, gametic chromosomes of *D. pur-purea*, 28 pairs; 3, the same of *D. ambigua*, also 28 pairs, with the hybrid between them (2) with 56 somatic chromosomes; 4–9, chromosome divisions in the F_1 gametes, showing the failure of the chromosomes to separate and form two separate cells, thus giving rise to gametes with the full somatic number of chromosomes (56) instead of the reduced number (28). From these arose in F_2 individuals with twice the number of chromosomes (112), giving 56 pairs of chromosomes and 56 chromosomes in the gametes (see 11 and 12), which were fertile, thus giving rise to a new constant species (*Digitalis mertonensis* Buxton and Darlington). (After Buxton and Newton.)

in a wild state. We have already mentioned a few of these which produced new fertile polyploid species from sterile diploid hybrids (fig. 120). We will now note some other types.

An interesting new plant species has been created experimentally by

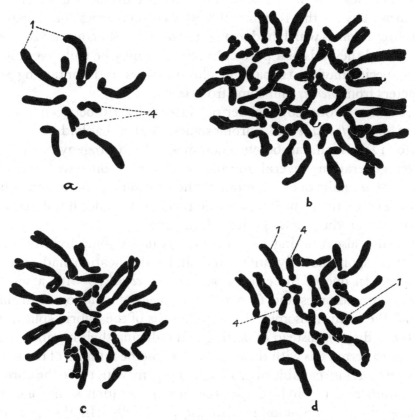

Fig. 121. The production of a new species in *Crepis* (Hawksbeard). *a*, the somatic chromosomes of *C. setosa* and, *b*, of *C. biennis*, the two parents with 8 and 40 chromosomes respectively; *c*, the F_1 hybrid with 24 chromosomes and, *d*, the constant and fertile species which arose from this in the fourth generation having 10 pairs of *biennis* chromosomes and 2 pairs of *setosa* (24 somatic); the pairs from *setosa* are marked as in *a*. (After Collins, Hollingshead and Avery.)

Babcock and his colleagues in California by crossing two species of the Hawksbeard (*Crepis*) (fig. 121). One species, *C. biennis*, is an octoploid, which has apparently arisen by reduplication, having 40 chromosomes, the other, *C. setosa*, has only 8 chromosomes. In crossing these together the hybrid has 24 chromosomes, 20 from *biennis* and 4 from *setosa*. When the germ-cells are formed the 20 *biennis* chromosomes pair together and form 10 pairs while the *setosa* chromosomes remain unpaired. In the fourth generation of this hybrid (some fertile seeds formed in each generation owing to the stability of the *biennis* chromosomes) a constant form arose having the 10 pairs of *biennis* chromosomes plus 2 pairs of *setosa* chromosomes which had segregated out after the constant breeding. This formed the basis of a pure culture which may be truly regarded as a new species having 12 pairs of chromosomes and containing some characters from *biennis* and some from *setosa*.

In *Nicotiana* (Tobacco) Webber produced a hybrid between a diploid gamete of *N. tabacum* (48 chromosomes) and a haploid *N. sylvestris* gamete (12), which had 60 chromosomes. Selfed progeny of this gave in later generations several constant new races, some with 24 pairs, others with 25 pairs of chromosomes, and containing characters of both species. These rank as new taxonomic races and would, if isolated from the parents, produce new species (cf. fig. 122).

Navashin, also working with *Crepis*, examined some hybrids which had arisen spontaneously and found that reduplication had occurred here also. In one of the crosses between *C. capillaris* with a set of 3 gametic chromosomes and *C. aspera* with a set of 4 he found in many cases all the 7 chromosomes in the germ-cells of the hybrid split in two as in the body-cells instead of pairing and reducing their number. These gave rise to germ-cells with the entire chromosome number of the hybrid and were therefore capable of producing progeny with twice the chromosome number of the hybrid. In the next generation some plants he examined had 10 chromosomes in the body-cells. In the germ-cells these formed 3 bivalent and 4 univalent chromosomes. Since the univalents split it was possible for the gametes to receive 7 chromosomes as in the first cross, thus providing another mechanism whereby constant and fertile species from the union of two unlike species may arise, only having double the sum of the chromosomes contained in the two original species.

An interesting case of the experimental reconstruction of a wild Linnean species by hybridisation occurred in the cultures of Müntzing

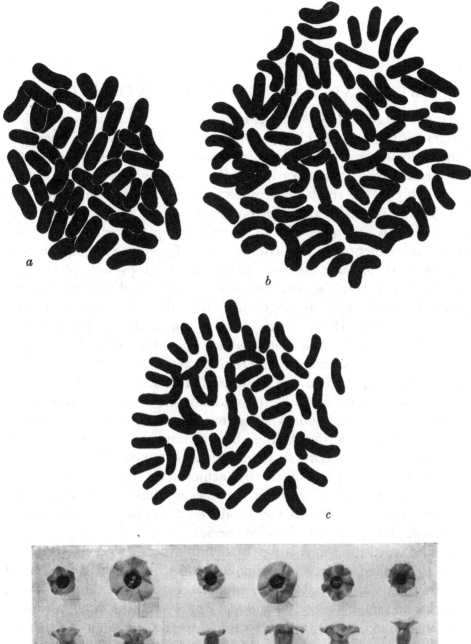

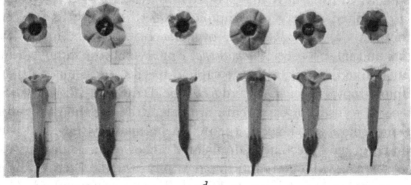

Fig. 122. Evidence from back-crosses shows that in the F_1 hybrids between *Nicotiana paniculata* (24 chromosomes) and *N. rustica* (48) 32 per cent. of the viable female gametes contain approximately the whole somatic number of chromosomes of the hybrid (36). As a few chromosomes are usually eliminated the progeny are very variable, owing to there being different chromosomes present or absent. *a*, a back-cross with *paniculata*, showing 46 chromosomes (2 chromosomes eliminated); *b*, another with 80 (in this case the gamete from the hybrid must have had nearly twice the somatic number); *c*, a back-cross with *rustica* with 59 chromosomes (only 1 chromosome eliminated); *d*, flowers of six of the back-crosses with *paniculata*, showing variation due to chromosome elimination. (After Lammerts.) (Cf. fig. 123).

(1930) in the genus *Galeopsis*. *G. pubescens* and *G. speciosa* are two diploid species with 16 somatic chromosomes and *G. Tetrahit* and *G. bifida* are two tetraploid species with 32 chromosomes. In an F_2 culture of *pubescens* crossed with *speciosa* a triploid plant arose with 24 chromosomes, presumably as the result of an unreduced gamete meeting a reduced one. This plant was highly sterile but produced one seed back-crossed to *pubescens*. This seed gave rise to a tetraploid plant, evidently by the failure in reduction in the triploid plant forming a triploid female gamete, with 24 chromosomes which, fertilised by a haploid male gamete with 8 chromosomes, produced the tetraploid with 32 chromosomes. This plant was indistinguishable from the pure tetraploid species *Tetrahit* (Common Hemp Nettle) both in appearance and genetically, crosses with *Tetrahit* and *bifida* (which are inter-fertile) being quite successful. This a particularly interesting case, showing how a Linnean species may have arisen by hybridisation between two other species followed by reduplication of the chromosomes. In the cultures the duplication arose in a somewhat roundabout way, taking two generations, and this is a further method by which it may take place, though probably the simple reduplication is more common. The synthesis of another Linnean species was also achieved in the Timothy Grass (*Phleum*) by the hybridisation of two species, and the subsequent reduplication of the chromosomes produced a regular polyploid species apparently identical with the wild species.

Many garden roses have arisen by hybridisation of a polyploid species with a polyploid variety of a diploid species. The Old Rose-scented Provence or Cabbage Rose, *R. centifolia* L., common in old-world gardens, and its sub-species *damascena*, from which the famous attar of roses is obtained, belong to the tetraploid species *AACC* made up of the two diploid septet species *AA* and *CC*. The Tea Roses (*R. odorata* Sw.), which came to us from China, are diploids with *AA* septets of chromosomes and characters (Hurst, 1929). In cultivation these *AA* sets of chromosomes have become reduplicated and we get some very large types of tea roses which are tetraploid and carry *AAAA* sets of chromosomes and genes (e.g. Gloire de Dijon). These two tetraploids (*AACC* and *AAAA*) crossed together gave a new type, the hybrid tea roses (e.g.

La France), which are tetraploids with the septet constitution *AAAC*. Thus the two parents, the tetraploid species *AACC* (gametes *AC*) and the tetraploid variety *AAAA* (gametes *AA*), gave a tetraploid hybrid *AAAC*. There is a good deal of sterility in these hybrids owing to the trivalent *A* and univalent *C* chromosomes, but in subsequent genera-

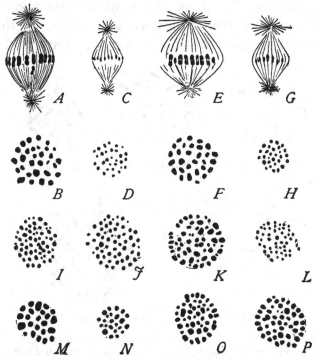

Fig. 123. A similar phenomenon in Moths in which Federley found that in hybrids the chromosomes divided instead of reducing in the gametic divisions, so that the gametes contained the full somatic number or nearly so. These on being back-crossed to the diploid parents gave triploids. *A, B, C, D, Pygaera anachoreta*, gametic divisions, 30 chromosomes; *E, F, G, H*, the same in *P. curtula*, 29 chromosomes; *I, J*, hybrid *anachoreta* ♀ × *curtula* ♂ with 59 and 58 chromosomes; *K*, back-cross *anachoreta* ♀ × (*anach.* ♀ × *curt.* ♂), 56 chromosomes and, *L*, later stage with about 50; *M, N, P. pigra*, 23 chromosomes; *O, P, pigra* ♀ × *curtula* ♂, 46 and 48 chromosomes. (Cf. fig. 122.) (After Federley.)

tions the chromosomes become balanced again with bivalents in new and more fertile recombinations. In these garden hybrids we get a mixture of allo- and auto-polyploidy, which may be distinguished as amphipolyploidy, although in the more complex hybrids these distinctions break down.

Crane and Darlington report a similar occurrence in blackberries, arising however in a different manner. Here a cross was made between a diploid species (*Rubus rusticanus inermis*) and a tetraploid species (*R. thyrsiger*). The resultant seedlings were all triploid with the exception of one which was a tetraploid. This had evidently arisen from a gamete of the diploid having failed to reduce its chromosome number and thus containing two septets instead of the normal one septet. On these meeting the two septets in the gamete of the tetraploid, a tetraploid hybrid was

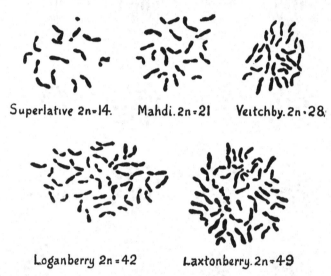

Superlative 2n=14.   Mahdi.2n=21   Veitchby.2n=28.

Loganberry 2n=42   Laxtonberry.2n=49

Fig. 124. Polyploid hybrids in *Rubus*. The heptaploid Laxtonberry with 49 chromosomes was derived from a cross between the hexaploid Loganberry (42) and the diploid Superlative Raspberry (14) and must have originated from a diploid female gamete in the same way that a heptaploid arose in the cultures of Crane and Darlington from a similar cross. The Veitchberry may have had a similar origin. (After Crane and Lawrence.)

the result instead of the expected triploid. The same authors, in crossing the loganberry (fig. 124), which is hexaploid, with a diploid raspberry, procured two heptaploid seedlings (with seven septets) as well as the expected tetraploids (with four septets). This was evidently another case of an unreduced gamete having been fertilised. The loganberry itself is believed, from genetic and cytological evidence, to have arisen as a duplication from a triploid raspberry-blackberry hybrid, thus forming a fertile true breeding and regular new hexaploid species. Soon after the introduction of the loganberry by Judge Logan of America as a hybrid between the blackberry and the raspberry, the author raised several hundreds of selfed seedlings, and since all bred specifically true, with no segregation of blackberry and raspberry characters, he concluded that it was not a hybrid as alleged but was probably an unknown wild species from the United States. This was long before the chromosomes of the loganberry were examined. Now it is evident that it was a hybrid after all, but it is also a true species experimentally created by hybridisation and the duplication of the chromosomes.

Bremer had a similar experience with eight hybrids raised from two distinct sugar cane crosses as found in the blackberry hybrids. In both cases the mother had evidently contributed a double set of chromosomes, but as there was some segregation of maternal characters, Bremer believed it probable that there was a long split of the female chromosomes during the first division of the embryo while the male chromosomes remained unsplit. In this case a high number of chromosomes is involved, the first five hybrids having 136 instead of the expected 96 (40 + 56) and the other three having 128 instead of 88 (40 from one parent and 48 from the other). These hybrids behaved fairly regularly during division and produced some good pollen.

Many of the polyploids which occur in cultivated plants are extremely complex in their constitution, especially roses and fruit trees, owing to many generations of hybridism and artificial selection, and they often contain very unlike sets of chromosomes, which have entered into their composition from the varied parents of past generations. In cultivation these heterogeneous sets can apparently persist in a manner which is often surprising, but one must consider that the conditions under which they live are all favourable to their survival and most of them are artificially propagated, which gives them a potential immortality without sexual reproduction and thus a chance to regulate themselves which would be denied under wild conditions.

In the Tomato, Lesley finds that on crossing triploids (which have arisen by crossing tetraploids with diploids) with diploids he gets numbers ranging from 24 (the normal diploid number) up to 27. Thus plants arise with one to three extra chromosomes, in the same way that they have arisen by non-disjunction of chromosomes in many cases (p. 88), and one is able to test the characters peculiar to each individual chromosome by different ones appearing in triplicate (fig. 125), nine of the expected types having been procured. *Oenothera* mutant types have also been obtained in this way. Under cultivation the plants usually return to normal conditions in later generations, but one would expect occasional individuals to fix a new type with one or more extra

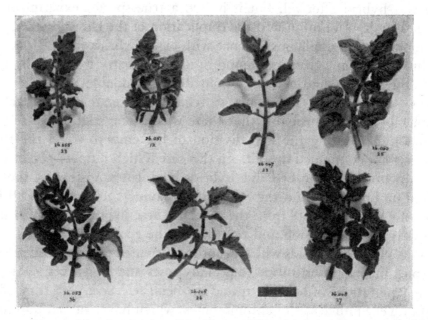

Fig. 125. Leaves of six trisomic types of the Tomato "Dwarf Aristocrat" compared with the normal diploid (bottom middle). Each of these plants has a different chromosome present in triplicate, giving a somatic chromosome number of 25 instead of the normal 24. From left to right—triplo-*F*, triplo-*D*, triplo-*A*, triplo-*B*, triplo-*C*, normal diploid and triplo-*E*. (After Lesley.)

pairs of chromosomes, since in hybrids at any rate odd chromosomes usually split equationally during the formation of the gametes, and thus a large percentage of the gametes may be expected to carry the extra chromosome or chromosomes if not lost in later stages.

Darlington (1928) finds that all the Sweet Cherries examined are hyperploid (i.e. have one or more extra chromosomes) and have from 17 to 19 somatic chromosomes instead of the normal 16. These extra chromosomes presumably carry genes which improve the individual from a horticultural point of view, since their occurrence in the cultivated varieties shows that forms containing them have been unconsciously selected by generations of raisers, who would naturally discard all those which were not an improvement on their general stock. Their continuance in this case is due to the custom of budding or grafting, since seedlings from them often return to the wild diploid state, the odd chromosomes being eliminated in the formation of the gametes; hence the seedlings are of no horticultural value.

Much work has been done with wild *Viola* species by Clausen with extraordinarily interesting results. In one cross between two nearly allied species, *tricolor* with 13 pairs of chromosomes and *arvensis* with 17, he gets various degrees of pairing in the first generation between 13 pairs and 4 singles and 15 pairs, the singles usually splitting at reduction. By their action he believes that *V. arvensis* has at one time arisen from *V. tricolor* by chromosome non-disjunction. In subsequent generations he obtained intermediate fixed types with intermediate chromosome numbers. One plant in the first generation, however, behaved differently, occasionally forming only six pairs at the reduction division. In these cases the remaining single chromosomes split, with the result that in the next generation a plant arose with a greatly increased chromosome number and in the following generation plants with a greater number still (43 and 45 chromosomes) arose. In these the germ-cell formation was not quite regular but more regular than in the preceding generation, and it is probable that the chromosome number will become constant when the single chromosomes become balanced in later generations and new fertile species arise with a much higher chromosome number. In both these generations the plants were strongly fertile and vigorous, with large flowers, in spite of the unbalanced chromosomes, due no doubt to the presence of one complete set of chromosomes and the two original species not being too unlike, natural hybrids between them being of frequent occurrence in districts where they overlap. In a state of nature *tricolor* thrives to the greatest advantage on acid soils while *arvensis* is found on basic ground, the hybrids being found between. Whether the difference in genetical constitution causes each

Fig. 126. Progeny of a triploid hybrid between a tetraploid variety of *Crepis capillaris* ($2n = 6$) and *C. tectorum* ($2n = 4$), showing the effect of varying numbers of *tectorum* chromosomes added to the normal $2n$ set of *capillaris*. Above to the left is the normal *capillaris*, in the middle and to the right three plants having $2n$ *capillaris* together with the D, B and A chromosomes of *tectorum* respectively. Below are four plants with the $2n$ *capillaris* complex with $A + B$, $A + D$, $A.+ D + C$ and $A + B + C$ *tectorum* chromosomes respectively. (After Hollingshead.)

one to be better adapted to its individual habitat is still an open question, but it seems significant.

It is also possible for new types of tetraploids to arise from triploids in much the same way, since in triploids (which may arise either from a diploid by the action of an unreduced germ-cell or from a cross be-

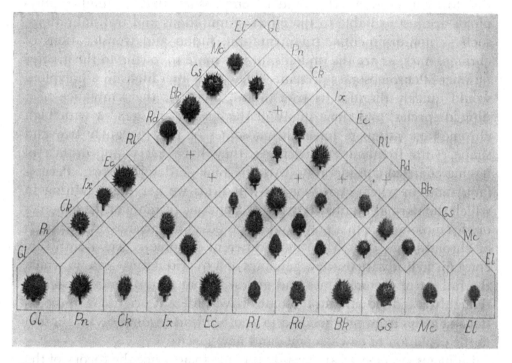

Fig. 127. Capsules of primary $(2n + 1)$ and double $(2n + 1 + 1)$ chromosome transmutant types in *Datura Stramonium* which have arisen from triploid cultures. Below are eleven of the possible twelve primary types and above are the double types which, since they have two different extra chromosomes, are really combinations of two primary types, the diagonal lines denoting the two chromosomes involved in each case. The crosses denote combinations which were raised but of which no capsules were available. (After Blakeslee.)

tween a diploid and a tetraploid) the unpaired chromosomes in some cases split at the reduction division, and if all are included in the germ-cells these carry two complete sets of chromosomes and may thus produce balanced tetraploids. This may not be achieved in the first generation but arise in later ones. If the triploid is the result of a cross this will give rise to a new type of tetraploid constitution.

Other balanced polyploids—hexaploids, octoploids, decaploids, etc.—may also arise in similar ways to those we have seen in the lower numbered types. Many types, however, occur through animals and plants which, not being multiples of lower numbers, must have arisen by irregular distribution of one or more chromosomes, probably largely due to hybridisation. It should be emphasised that the higher polyploids are just as liable to the effect of mutations and transmutations, such as non-disjunction, fragmentation, fusion and translocations of chromosomes, as are the diploids, in fact more so, owing to the greater number of chromosomes present. A few such alterations in a polyploid would quickly disguise its true origin, giving it the semblance of a diploid species, and since many of the species and genera with high chromosome numbers have chromosomes of very irregular size and shape, with frequent constrictions, they have every appearance of having changed considerably. Also in some garden roses (e.g. Pernetiana) and in wheat and wheat-rye hybrids we get the condition in which the normal number is preserved but one or more of the genomes or chromosome sets in a polyploid is composed of chromosomes or parts of chromosomes of two or more different genomes. Apparently, by constant hybridisations, new genomes of a hybrid nature arise in which the full set of chromosomes and genes normal to the genus is present though the individual chromosomes have different origins. Translocations of genes from one genome to another lead through hybridisation to the evolution of new genomes and new species.

Linnaeus (1744) was apparently the first to advance the theory of the origin of species by hybridisation and he was followed by Herbert(1820), Naudin (1852), Kerner (1871), Hurst (1898) and Lotsy (1912).

During the last decade the experimental creation of new polyploid species under the strict control of genetical experiments has provided conclusive evidence in support of the theory. There can be no doubt that the bringing together of many diverse genes by hybridisation, and their various interactions producing different results from their behaviour when separate, have been important factors in promoting the increase of complexity found in the evolution of plants and animals.

CHAPTER XI

GENE MUTATIONS

WE have examined the changes which have been brought about in animals and plants by the various transmutations of chromosomes, i.e. by the different combinations, additions or losses of sets of chromosomes, whole chromosomes or parts of chromosomes or by the duplication of entire sets, especially when the individual is the result of a cross between distinct species or genera, although these transmutations are of vital importance in creative evolution, it is obvious that all the changes in characters necessary to produce the divergent forms found in nature could not have arisen originally in this way alone. All these different forms of life could not have arisen simply from a reshuffling and recombination of the chromosomes and their genes, however often this may have taken place or however varied its character. This alone is not sufficient, there must also have been changes in the genes which go to make up the chromosomes, since the genes are the basic units of inheritance and evolution, the chromosomes being the important structures or carriers which bind the genes together and enable them to act as a functional whole.

Darwin (1859) concluded that all plants and animals vary slightly, almost imperceptibly and gradually in all directions, and that those changes which were favourable to the plant or animal were retained while those which were not favourable died out by elimination or Natural Selection. He was also aware of large and sudden jumps of variation, which he looked upon as monstrosities (and which they very often were), and he believed that these could play little part in evolution. De Vries (1900), experimenting with the Evening Primrose (*Oenothera*), found in his cultures plants that suddenly changed considerably and these he called mutations. We have seen (p. 86 and p. 100) how a recent examination of the chromosomes has demonstrated that the origin of most of these is due to transmutations of chromosomes, but there are others which are the result of gene mutations.

Morgan (1925), in his *Drosophila* cultures, found that certain marked changes took place in many characters, for instance, a red-eyed race would suddenly give off white-eyed individuals, and these were regarded as evidence that there was a constant succession of sudden and fairly large mutations taking place and giving ever fresh forms. So great, however, has been the number of individuals of this fly bred, and so minute the examinations to which they have been subjected, that it has now been found that extremely slight heritable changes, almost

Fig. 128. Different types of wings arising by mutation in *Drosophila melanogaster* arranged in order. *a*, cut; *b*, beaded; *c*, stumpy; *d*, another stumpy; *e*, vestigial; *f*, apterous. (After Morgan.)

imperceptible, may also take place. In the case of eye colour, although the very marked change of the red to the white eyes takes place occasionally, in addition many more minute changes take place, so that now numerous races have been formed which carry every grade of eye colour between the red and the white. In the same way long-winged flies suddenly produced a race of flies without wings at all, but now many intergrades have also arisen (fig. 128). Many physiological changes have taken place, but many so slight that only the most intimate acquaintance with the form of the creature and the immense numbers worked with have made them perceptible. From the original species of *Drosophila melanogaster* hundreds of different races have originated, some of them so different in their combinations of mutations that

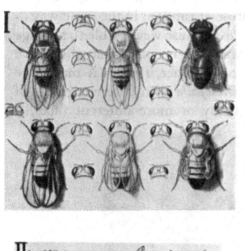

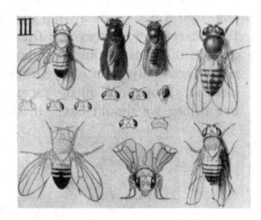

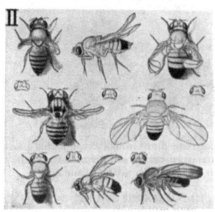

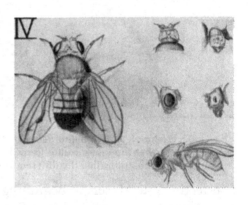

Fig. 129. Mutations which have arisen in each of the four chromosomes of the Fruit Fly (*Drosophila*). I, some which have arisen in the *X* chromosome, eye colours, wing, leg or bristle characters; II, in one of the autosomes, other eye colours, wing, leg and bristle characters; III, other modifications carried by another autosome; IV, those carried by the small autosome. (After Morgan.)

if found in a wild state they would certainly be classified by systematists as distinct species or even genera (fig. 129).

In a fossil bed in France a perfect series of the mollusc *Paludina* was found in which one type graded into another one quite different, during long ages of time. This was taken as a definite proof of the slow change by infinitely gradual steps from the one to the other, but since in the central layers many of the grades were more or less found together (fig. 130) it is more likely that they arose at random in the same way as the different coloured eyes in *Drosophila*. Because we are able to arrange a series of forms in this continuous way it is no proof that they have arisen gradually from one extreme to the other, though it may have been so. If one took all the races of *Drosophila* with different grades of eye colour we should have, in the thirty or more different races, a

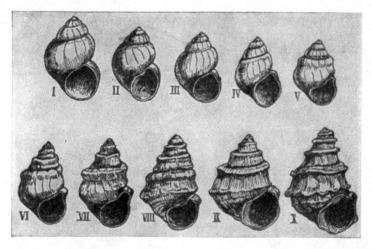

Fig. 130. Evolution of the shell in *Paludina*. The two extreme forms would be considered good species except that the intermediate forms lived contemporaneously with them and probably arose as mutations like the different eye colours in *Drosophila*. (After Neumayr.)

perfectly graded series beginning with the darkest colour, which is deep brown, to pure white. This, however, would give an entirely false idea of the actual happenings. As we have seen, the simple white eye first arose in experimental cultures, straight from the complex wild red, and it was only in later cultures that the intermediate and darker shades arose in various ways and in no definite order.

Thus it is possible to get these hereditary changes, or gene mutations as they are now known, in various degrees. Some involve quite a leap from one distinct thing to another. We are all familiar with cases of flowers, for instance, which, although usually deeply coloured, have white varieties or varieties of another colour. We have most of us searched for a plant of white heather when visiting a heather moor. On the other hand it is equally possible to get gene mutations which involve such a small change from the original type that they are all but imperceptible to the human eye, as the author found in his experiments with slight shades of scarlet flowers in *Antirrhinum* (1925), and it is highly probable that these small mutations have been an equally potent factor in evolution as the larger ones, especially in the early days of evolution when organisms were more simple in structure.

The modern study of genetics has been closely bound up with the origin of new characters by gene mutations. The elementary principles of Mendel's Laws could not in fact have been worked out at all if there had been no differential characters to work with. In the early days plants such as peas, which had already produced various mutations, were used, but later, when more work had been done, many new mutant characters arose suddenly in the various plants and animals used in experiments, and it was found more useful to work with those which had occurred in pedigree cultures and whose origin was known with certainty. These new mutant characters arise quite suddenly and maintain themselves with the same constancy as the characters of the original type from which they arose. They are more often recessive to the characters of the original type (fig. 132), but many cases of dominant characters arising by mutation of the genes are known. In either case, if they are fitted to survive at all, races will soon arise which will breed true, since either with a recessive or a dominant gene one individual out of every four will have the new character in a pure state in the second generation if inbred, and one out of two if back-crossed with the recessive parent (fig. 13). When a step has been taken in a certain direction it often happens that other mutations arise carrying the race on a further step in the same direction, not so much because the new type is more likely to mutate in that direction, as that individuals which have mutated in that direction, if it be an advantageous one, are more likely to survive. It is apparent also that some genes are much more unstable in one direction than others which appear to be more or less unstable in many directions.

From time to time curious mutations arise which, when present in a pure state (homozygous), cause the death of the animal or plant. Perhaps the best-known instance of this is yellow coat colour in mice (fig. 131). This is due to a dominant gene, and when only one dose of it is present we get mice with yellow coats which are apparently healthy, with the exception of a tendency to be over-fat. When these are bred from, however, instead of getting three yellows (one pure dominant and two impure dominants) to one pure breeding recessive brown as expected, we get only two yellows to one brown. On breeding from these

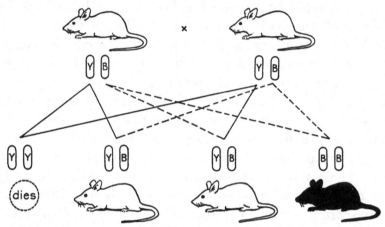

Fig. 131. The inheritance of yellow coat colour, a lethal gene in Mice. On crossing two yellow mice they throw one agouti brown (showing that they are heterozygous) to two heterozygous yellows like themselves, giving a 2 : 1 ratio instead of the expected 3 : 1. The pure yellows never appear, showing that a double dose of the gene is fatal and individuals of this constitution do not develop. (After Crew.)

yellows we find that they are all impure dominants carrying one gene for yellow and another for brown, all the pure dominant yellows having died before birth, the presence of the two genes for yellow causing them to be incapable of development. Many such genes have been found in cultures of various animals and plants and are termed "lethals", owing to their death-producing qualities. If they occur in the sex chromosomes the result is even more disastrous. In *Drosophila*, for instance, a recessive lethal present in the X chromosome will cause the death of all males of certain races owing to the absence of dominant characters in the Y chromosome to mask it, while in some of the females the presence of a covering dominant in the other X chromosome will prevent it from appearing. This may explain the fact that in certain species with XY chromosomes males are more delicately constituted than the females of the same species and more sterile as hybrids, while conversely in those organisms which have the ZW type of sex chromosome the female is often the more fragile sex and Haldane finds them more sterile as hybrids. In any case the presence of sex-linked lethals upsets the normal sex ratio of one-half males and one-half females in the same way that the normal three to one ratio is upset when the lethal genes are present in the autosomes. Although they may have small significance from an evolutionary point of view, except in so far as they produce sterility barriers between closely related species, yet the lethal mutations have a peculiar interest as a further demonstration of the fact that many mutations arise at random and not as a direct result of the influence of the environment.

Many of the fluctuations in characters or acquired modifications, which frequently occur but which are only potentially inherited, are due to the fact that many characters are not controlled by one gene alone but by several, so that the character is not influenced so much by the direct action of a single gene but by the balance of the interaction of all the genes with which it is concerned. Conversely, many genes exist which influence more than one character of an individual, so that if that gene changes in any way then all the characters which it influences are also affected. How the genes effect these changes is not yet clearly understood, and we have not yet learned the precise nature of the gene or its mode of action. We have already observed the normal extraordinary stability of some genes, how the green colour in peas preserves its identity although masked by yellow for many generations. Similarly, if we cross a pure bred coloured rabbit with an albino, although the cross-breds are all coloured they produce gametes with genes representing coloured or white in equal numbers and one in every four of the second generation will be pure white. No matter how long the genes for albinism lie dormant in the cells of the cross-bred coloured rabbits the coloured hair of the cross-bred has no influence on the germ-cells, and the genes representing white remain uncontaminated ready to develop albinism whenever two come together in fertilisation.

Some genera and families seem to have a much greater gene stability than others. Among animals we find various large groups, such as the Lampshells (*Brachiopoda*), which have gone on and on through long ages

Fig. 132. On crossing a White Leghorn Cock (*A*) with a White Plymouth Rock Hen, (*C*) the F_1 is white with a little flecking of black (*B*), but in F_2 barred birds appear as in *D* showing how genes for colour and pattern may lie dormant in apparently pure white individuals. In this case crossing the two different breeds brings together the hidden recessives necessary to produce the colour and pattern. (After Hadley.)

of time showing only minor variations in their characters. In the plant world we also find several similar cases where evolution seems to have stood still for ages, as, for instance, the Sea Weed *Laminaria* and the Red-Wood Trees of California (*Sequoia*).

Recent work by Sax on Rhododendrons gives an excellent demonstration of genetic stability in a large genus. Eastern Asia was evidently the original home of the rhododendrons and from here they appear to have migrated to America, where many Linnean species are found, by way of the land connection between the two continents which existed up to the time of the Pleistocene Ice Age (fig. 133). Since then communication has been cut off, yet an examination of the chromosomes of the American and Asiatic Linnean species shows that their chromosomes are similar (13 being the gametic number) not only in outward appearance but also in specific gene content, since it is possible to raise fertile hybrids between them, and in these hybrids the chromosomes show

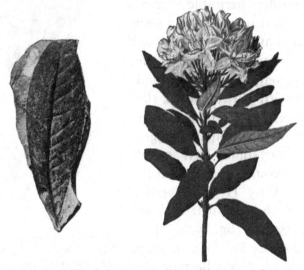

Fig. 133. On the left a fossil Rhododendron leaf from Pleistocene deposits found in the Inn Valley, on the right a flowering branch of a modern Rhododendron species, showing their great similarity. (After Peake and Fleure.)

every sign of compatibility, pairing and giving regular reductions in the formation of the gametes. Since it is probable that no communication has been possible between these Linnean species (or genetical sub-species as one may more properly call them from this genetical evidence) for hundreds and thousands of years, this compatibility between them is a remarkable case of the stability of the genes, which have remained unchanged for vast periods of time. Similar instances are found in *Rosa* (Hurst, 1928), in which a large group of Linnean species formerly considered as a section of the genus have had to be brought together, owing to the experimental evidence that they belong to one large genetical species with many geographical sub-species, which, in spite of long isolation from one another in Europe, Asia and America, still show complete specific stability of their chromosome complements and gene complexes. The sub-specific and varietal differences which have arisen have not caused any incompatibility, so that completely fertile hybrids are possible between European, Asiatic and American members of the group.

That genes change, however, is evident, or the sudden heritable changes or mutations could not arise. In some cases it is clear that the germ-cells have been affected by some treatment. Various experiments have been made and treatment with X-rays (see p. 189) and various chemicals and temperatures have affected them to such a degree that heritable variations have arisen. Recently an interesting experiment has been carried out by Heslop-Harrison and Garrett on the Geometrid Moths (*Selenia* and *Tephrosia*) (fig. 134). Many moths have what are known as melanic varieties, that is, they have become very dark in colour. Since this has occurred principally near large cities, an idea arose that it was due to the chemicals

Fig. 134. Below, the normal variety of the moth *Selenia bilunaria*, and above, the melanic variety, a mutation apparently induced by feeding the caterpillars of the previous generation on foliage treated with manganese salts. (After Harrison and Garrett.)

deposited on the vegetation from the smoke given off from the various industrial works. To test this, Heslop-Harrison took some normal light-coloured moths free from melanics and fed their larvae on artificially infected foliage, and succeeded in raising a melanic race in the second generation which bred true to colour when normally fed. In this experiment it appears that a single gene in the germ-cells was changed by the chemicals introduced by the food of the animal, and it is possible that other mutations may be caused in this manner. Extremes of temperature, climate, variation in soils and foods, and various conditions of life, may each have an effect upon the genes of organisms and cause them to mutate in many directions. If it is a favourable change in the right direction the organism will be better fitted to survive in changed conditions, and the new race will establish itself and replace the old type, which may not be so well fitted for the new conditions.

Many of the mutations which have arisen in *Drosophila* are purely pathological and could hardly survive in a wild state. The wingless mutants which arose, for instance, might be regarded as pathological and unable to thrive naturally, but on the other hand in some species of moths only the males are winged, the females being wingless, spending the whole of their existence in approximately the same place in which they originated. In ants, also, some forms are winged and others wingless, so the wingless state appears to be a not uncommon mutant in nature. Other mutations which have arisen in cultures, such as different eye colours and structural features, are also found in wild species in their normal environment (fig. 135). Lethal mutations are naturally pathological, since the individuals carrying them in a homozygous state die, but it is probable that many of these occur in nature unobserved, since they may persist indefinitely in a heterozygous condition when recessive, or, in many cases where the conjunction of two different genes is necessary to produce the lethal effect, they may persist in different species and only come together when rare crossings occur, when they are eliminated by the death of the offspring.

With regard to the possible effect of environment, the *Drosophila* cultures are kept under standard conditions and their food is usually a uniform diet, so that there can be little cause for induced mutations due to changes in the conditions of life. Incidentally, it is difficult to associate many of the mutations with environmental influences at all, since so far as we know they cannot be considered to have any effect whatever on the survival of the insect. On the other hand, some mutations have occurred occasionally in cultures of other plants and animals through which new strains have arisen capable of living at different temperatures or under different conditions of life. The Florida velvet bean, which originally could only be grown in Florida and the Gulf States of America, suddenly produced a mutation capable of growing and setting seed in more diverse climates and thus capable of being cultivated throughout a much larger area of the United States. In *Drosophila*, however, few of the mutants can be regarded as being in any way either better or less adapted to their environment, and one must apparently look on these mutations as purely random and indeter-

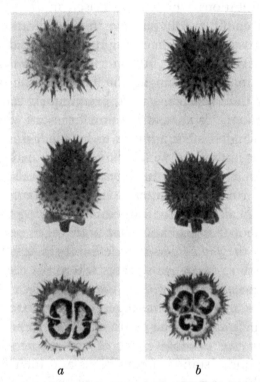

a   *b*

Fig. 135. A gene mutation arising in the Jimson Weed (*Datura Stramonium*). This plant has normally (*a*) two carpels giving two main divisions in the capsules while the mutant (*b*) has three carpels and is known as "Tricarpel". The two-celled ovary of *Solanaceae* is of high taxonomic value as a family character and its mutation to a three-celled ovary creates a link with the family *Polemoniaceae*, which belongs to a different sub-order. (After Blakeslee, Morrison and Avery, *Journal of Heredity*.)

minate variations. Recent work on the effect of X-rays throws much light on mutations of this order and their causes (p. 189 and p. 200). That many random mutations arise in nature by the direct influence of the environment on the germ-cells can hardly be doubted, but the view that all mutations are due to direct environmental causes, the alteration fitting the creature better for that particular environment, cannot be held in face of all the evidence for random mutations which has been accumulated, although naturally many of the random mutations do happen to increase the survival value of their possessors in changed conditions, since those which do not are eliminated by natural selection.

Much work remains to be done and many experiments carried out before we can come to a full understanding of how the genes are affected by internal and external conditions. Some interesting work has been done by Julian Huxley, Ford and others on the rates of development of certain characters. Much might be learned of the genes if we could get an insight into how they work and form the characters, structures and functions to which they give rise. Many functions and parts of the body are controlled by the definite action of certain substances known as enzymes and hormones, each of which produces a particular biochemical reaction. That these are primarily due to the action of special genes can hardly be doubted and their investigation will no doubt help us to understand how other characters are developed (fig. 136).

The precise nature of the gene is at present not fully understood, but it is assumed to be some form of organic molecule or group of molecules. According to recent calculations the minimum size of the gene of *Drosophila* is estimated to be about the same as the maximum size of the largest protein molecule. With the conception that the gene is a complex organic molecule of this type comes the idea that it is the ultimate basis of all life—that the primary gene or protogene was, in fact, the first manifestation of life, growth and reproduction. That the genes have the power of growth is evident from constant observations that the chromosomes and their genes are continually growing and dividing and forming new cells and new bodies, and it is a natural inference that the origin of the gene is coeval with the origin of growth, development, reproduction and of life itself, with all its attendant properties and values. With our present knowledge it is inconceivable that the complicated organisation of the cells and their contents could have come into being without the existence of genes to organise and control the development. In short, we conclude that the gene is the organiser and the foundation of life. Although as yet we know very little about the structure of the gene and how it mutates and organises development, the fact cannot be obscured that the genes have been experimentally demonstrated in

Fig. 136. Examples of short-leggedness in cattle occurring on several American ranches. It behaves as a simple Mendelian dominant character but is believed to be due to the under-functioning of the pituitary gland (which is unusually small in affected animals), thus showing that the abnormal functioning of glands, often affecting several regions of the body, may be as truly genic and inherited as an ordinary bodily character, and it is highly probable that many mutations are of this type. (After Lush, *Journal of Heredity*.)

thousands of experiments by hundreds of investigators in all the forms of life that have been examined. The genes prove to be the organisers

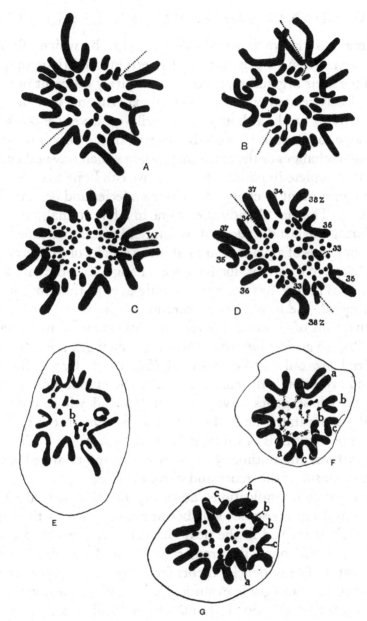

Fig. 137. Chromosomes of the domestic fowl: *A*, female with one large sex chromosome, *B*, male with two (♀ $2n = \pm 35$, ♂ $2n = \pm 36$); of the Indian runner duck: *C*, female with one large sex chromosome, *D*, male with two (♀ $2n = 77$, ♂ $2n = 78$); and of the reptiles: *E*, the tree lizard *Sceloporus spinosus* (♂ $2n = 22$), *F*, *Sceloporus undulatus* var. *corsobunus*, a rare lizard (♂ $2n = \pm 30$), *G*, the American chameleon *Anolis carolinensis* (♂ $2n = 34$), showing the striking similarity of shape and arrangement of chromosomes and number of large chromosomes in birds and reptiles. (After Hance, Werner and Painter.)

of the hereditary characters and the sole means by which life is carried on from generation to generation. In certain conditions they mutate and change their organisation, not to an entirely different organisation but to a closely related one. In genetical experiments we can see evolution actually at work, although necessarily on a very small scale. We cannot hope to produce in our laboratories, in the very limited time at our disposal, changes of the same magnitude which have taken millions of years to complete in nature. The most we can hope to do is to induce an animal or a plant to change into a new species and occasionally into a new genus. This has already been done in a few animals, and in many plants during the last decade, but we cannot hope in our time to change a species or genus of a plant or animal into a species of a different family, order or class. We cannot, for instance, change a reptile into a bird or a mammal as nature has done in the long ages of time that have been at her disposal. We have in our experiments, however, seen enough of the mechanisms of chromosome transmutations and gene mutations to realise that, given secular time, such great changes are quite possible and indeed are only to be expected (fig. 137). By far the greatest mystery in evolution is the continued existence of numerous species and even phyla whose genes have not mutated and whose chromosomes have not transmutated and whose characters remain the same to-day as they were thousands of millions of years ago. Their fossils remain a standing witness and testimony to the stability and relative immortality of the mechanism of the genes and chromosomes.

There are definite indications, however, that this natural stability of the genes and chromosomes will become more rare as time goes on. Recent work, described in the next chapter, has given man the power to speed up mutations with great rapidity, and one may confidently look forward to future developments on a large scale of the capabilities of man to control and guide evolution at his will by the creation of new species and the substitution of human for natural selection. In view of the recent discovery of a genetical formula for the inheritance of intelligence in man (Hurst, 1932), it is even possible to predict human control of the development and evolution of mind.

CHAPTER XII

EXPERIMENTAL MUTATIONS AND TRANSMUTATIONS

THE science of Genetics took another great step forward in the year 1927, when Muller in America announced that he had been able to produce by means of X-rays the same mutations in *Drosophila* that have arisen from time to time in the experimental cultures of Morgan and his colleagues, the only difference being that they occurred with much greater frequency. The use of X-rays in medical science has been largely and increasingly exercised for many years, and experiments have also been made on various tissues, both plant and animal, which showed that many abnormalities might arise after the penetration of the rays. Muller, however, conceived the idea of treating his living cultures of *Drosophila* with the rays, and his results created so much interest and enthusiasm that numerous workers are now repeating his experiments on different animals and plants.

It has been found that two main types of genetical modification may arise from X-radiations: (1) a new distribution of the chromosomes or parts of chromosomes producing new combinations of characters in new linkages, which, by altering the genic balance, gives rise to transmutations; (2) changes in single genes producing mutations.

In *Drosophila melanogaster* a very large number of gene mutations and transmutations have been induced by this means. These are identical with those which have occurred naturally in the previous genetical cultures without the aid of X-rays (fig. 138). The untreated controls did not mutate. The flies were treated with a suitable dosage of the rays at varying times of their development and the different results examined. In some cases the eggs or young larvae have been treated and in this case somatic mutations, or changes in the body-cells, have been produced, showing how some cells may be affected and others not. Patterson finds that if eggs and larvae derived from a cross between a normal

red-eyed female and a white-eyed mutant male were irradiated at
various stages of development, a proportion of the resulting flies, instead
of having red eyes (red being dominant to white), showed white areas
or patches in their eyes (fig. 139). A most interesting fact observed in
regard to these patches was that the size of the areas varied with the
period at which the rays had been administered. If the eggs were treated
within the first few hours of development the white patch would be very
large, but the later the treatment the smaller became the patch, and if

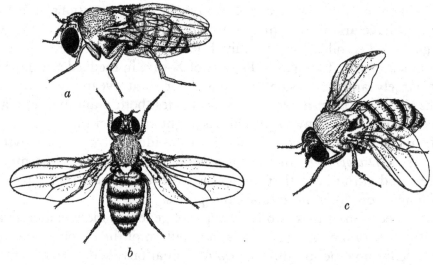

Fig. 138. Some effects of X-rays on the germ-cells of *Drosophila*. *a*, a dominant gene mutation,
prickly bristles, lethal in a pure state, located near rough eyes in chromosome III; *b*, a recessive
gene mutation, outstretched wings, at locus 62·5 in the *X* chromosome; *c*, a transmutation,
"swooping" wings, caused by an interchange of segments between chromosomes II and III,
lethal when homozygous. (After Muller.)

the radiation was performed in the pupal stage no white areas arose at
all, the emergent fly having pure red eyes. From this it is inferred that
the change took place in a single cell of the rudimentary eye, otherwise
it is difficult to explain the variation in size with age. In the very early
stages when the area of the eye is limited to only two or three cells, the
mutation in one cell which, subsequently dividing and subdividing,
forms a large area of the adult eye, causes the whole area of the eye resulting from this cell to consist of cells carrying the mutant gene which

arose in the one primary cell. It is evident that as the number of cells in the eye increases with age the later treatments with the rays will have far less effect, as the cell affected by them will influence a much smaller section of the eye, till, finally, when the eye is fully formed, the radiation will be powerless to cause any result at all.

The most powerful influence of the rays results from the radiation of adult flies, both male and female, before fertilisation has taken place. In this case the genes in the chromosomes of the germ-cells are affected, and when fertilisation takes place the resultant flies show the effect of the gene mutation throughout the entire part of the body which is controlled by that particular gene, and the change is inherited.

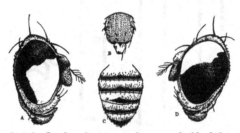

Fig. 139. *A*, right eye of male fly showing more than one-half of the eye white as a result of a somatic gene mutation, from a culture of eggs X-rayed at 7 to 8 hours old; *B*, thorax of a female fly showing left half "singed", from culture of eggs rayed at 1 to 7 hours; *C*, spineless area on abdomen at *X*, from a culture of eggs rayed at 1 to 10 hours; *D*, left eye of male fly with less than half the eye mutated from red to white, from a culture of eggs rayed at 14 to 16 hours old. (After Patterson, *Journal of Heredity*.)

More frequent than the change of individual genes under the influence of the rays is the marked tendency in treated material for a break to occur in one or more places of the chromosomes, leaving one or more fragments detached from their parental chromosome. In many cases these detached pieces join themselves on to one of the other chromosomes and we get a condition the same as found in the normal cultures in which translocations have taken place (p. 68).

If these abnormal individuals are crossed with normal ones we get this particular set of genes appearing in triplicate in one-half of the progeny, which causes a considerable variation in their appearance, while in part of the next generation it will be present four times and a constant race arise with the two longer chromosomes. The evidence of

these breaks has been very fully investigated genetically and cytologically by Muller and Painter. Fig. 140 shows an illustration of this.
In this case part of the third chromosome has become detached from its
own chromosome and joined on to one end of the second. Genetically,
it was proved that this particular group of genes had become free from
its usual companions and joined those of the second chromosome, the
exact end of the second chromosome also being proved by the association of the characters. Cytological examination showed that this was

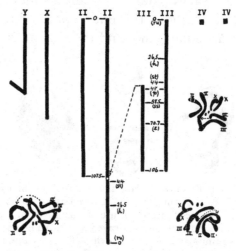

Fig. 140. Translocations caused by X-radiation. Genetical evidence showed that a large piece
was missing from one of the 3rd chromosomes, the break occurring between scarlet (*st*) and
pink (*p*), and that it had become attached to the lower "right hand" end of one of the 2nd
chromosomes, thus forming a new linkage group. Cytological evidence confirmed this, nearly
the whole arm of one of the 3rd chromosomes being missing and one of the 2nd chromosomes
being much longer than normal. (After Painter and Muller, *Journal of Heredity*.)

indeed the case, one of the third chromosomes being appreciably shorter
than the other, while one of the second chromosomes had become much
longer. This is a further demonstration that the genes which are represented as lying alongside in the chromosome maps actually do lie in
that order (fig. 30).

That this translocation of genes or transmutation is a potent factor
in evolution has already been pointed out. We have seen (p. 71) that
different species of *Drosophila* and other animals and plants have parts

of their chromosomes in different positions relatively to one another, and numerous cases have arisen in the cultures of duplications, losses and translocations of sections of chromosomes (figs. 140–142). Painter,

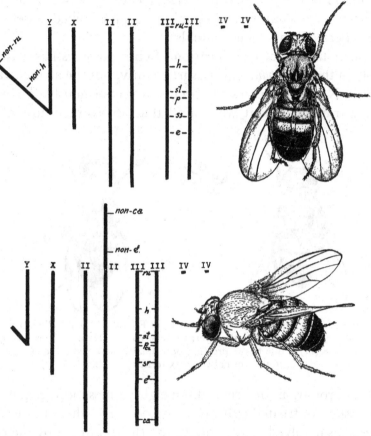

Fig. 141. Above, hyperploid transmutant Fruit Fly containing an extra segment from left end of chromosome III attached to *Y* chromosome and two normal chromosomes III. Differing from normal in broad convex wings, imperfect cross veins, reduced bristles, dark patterned thorax, misshapen eyes, a tendency to incurved hind legs. Below, hyperploid transmutant Fruit Fly with extra segment from right end of chromosome III attached to chromosome II, in addition to two normal chromosomes III. Differing from normal in divergent, slightly raised wings with imperfect longitudinal veins, long bristles, rather small eyes. (After Muller.)

working on the chromosomes of the Rat and the Mouse (two species of one genus *Mus*), found that while the chromosome number is nearly the same (42 and 40, respectively), and the amount of chromatin about the same, yet the distribution of the chromatin was very different, pointing out that as the rat and mouse have most probably arisen from a common ancestor, the change of species has been brought about largely by the redistribution of the chromatin in translocations between the different chromosomes. Similar instances might be multiplied indefinitely in various genera of plants and animals.

This phenomenon of translocation of chromosomes has arisen spontaneously in the *Drosophila* cultures frequently, but the artificial application of the X-rays increases the occurrence enormously. Thus for one particular translocation it was found that an average dose of X-rays

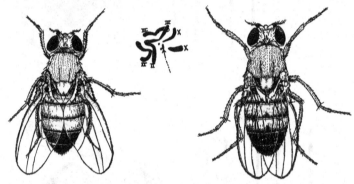

Fig. 142. On the left a hyperploid transmutant Fruit Fly called "Tubby" with a marked increase in width and decrease of length, divergent wings and bulgy eyes. This divergence from the normal type (right) is due to the presence of a small unattached piece of chromosome in addition to the normal complex (centre). (After Muller.)

caused the chromosome to break about eighty times more frequently than was the case in untreated cultures. Also, in cases where more than one character was involved, it was found that the treatment by the rays had no effect on the ratio between the two types of variation, but only on the increased frequency of their occurrence.

Whiting has tried similar experiments on Parasitic Wasps, but in this case it was found that much stronger doses of the X-rays were needed to produce any change at all, and this caused a great deal of sterility, and in some cases complete infertility. Only one definite mutation was induced, and it is evident that in this case the germ plasm is relatively more stable, which is interesting in view of the many forms of life which have remained little changed throughout long periods of general evolution, while others have changed relatively rapidly out of all semblance to their former selves.

With regard to plants, numerous experiments of a similar nature have

Fig. 143. Three Tobacco Plants (*Nicotiana tabacum* var. *purpurea*) raised from an untreated seed-plant fertilised by pollen from an X-rayed individual. The centre plant is much larger and more vigorous than the normal while the two others are very dwarf. All were fertile. (After Goodspeed, *Journal of Heredity*.)

been carried out. Much has been done in the Thorn Apple (*Datura*), and Goodspeed, working with the Tobacco Plant (*Nicotiana*) (fig. 143), has found that germ-cells treated by X-rays showed numerous abnormalities, the chromosomes fragmenting or going the wrong way during divisions. In this case the fragments, instead of going on to other chromosomes, more often persist as they are, thus giving a different number of chromosomes in subsequent generations. Another im-

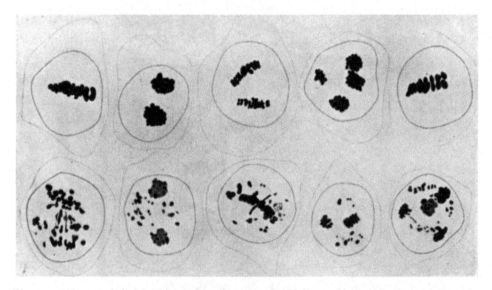

Fig. 144. Abnormal divisions in male gametes in the Tobacco Plant (*Nicotiana tabacum* var. *purpurea*) after treatment by X-rays. Above, normal divisions in untreated plants, below, abnormal divisions in treated plants, showing irregular distribution of chromosomes in the middle cell of the bottom row and the presence of only one spindle where there should have been two, which might give rise to a diploid gamete. (After Goodspeed, *Journal of Heredity*.)

portant occurrence was the production of gametes with twice the number of chromosomes (fig. 144) (see p. 100), which in the following generation would give rise to polyploids. It is perhaps significant that those species which show the greatest variation normally react to the treatment much more than other species which show relatively slight variations and are evidently more stable in their germ-plasm. Species with larger chromosome numbers were also more affected than those with fewer chromosomes. The ripe seeds showed themselves extremely resistant to any effects of the rays, it being difficult to find a dose strong enough to affect them. The same is true for many other genera. During these experiments three stable lines have been produced in *Nicotiana tabacum*. Triploids and tetraploids also arose.

Navashin, working with *Crepis*, found the average mutation rate increased 600 times but obtained rather different results. He subjected his material to X-rays for a total time of 14 minutes, though he did not keep it under the rays continuously but administered fractional dosages which amounted to this length of time. He found 100 per cent. abnormality in the seedlings raised, all of them showing a marked disturbance in the somatic chromosomes. The root tips were usually chimaeral, that is, they showed different chromosome complexes in different parts in the same way that the *Drosophila* eyes were affected. The majority of the abnormalities took the form of fragmentations and dislocations, usually associated, no polyploids being seen in these cultures. Curiously enough, the normal number of the chromosomes was not changed, only the relative lengths, since the fragments always joined on to one or other of the chromosomes. The same conditions are found in untreated plants, but comparatively rarely. Since the majority of *Crepis* species differ in the comparative length of their chromosomes, and evolution in this genus has evidently proceeded largely by differences in the arrangements of the gene complexes, it is interesting to find this type of transmutation occurring in the treated material, and it seems to show that each genus has its own particular type of instability and consequently its own peculiar mode of evolution. With this type of transmutation we get different characters much accentuated in different species, owing to the duplication of some parts of the chromosomes.

In maize eight specific characters, which had hitherto remained entirely constant, mutated under the influence of X-rays, thus providing valuable evidence that specific characters are also represented by genes (see p. 219). In wheats X-rayed by Stadler no variations arose, which he attributes to their polyploid condition, which causes recessive mutants to lie masked by the similar genes in the other sets of chromosomes predominating over them. One advance on the type, however, was induced by Delaunay, since it ripened four days earlier and was

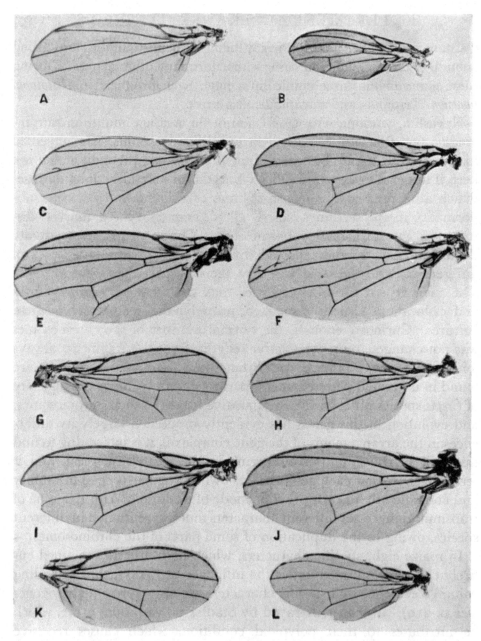

Fig. 145. Mutations induced by X-rays in wing characters in the Fruit Fly (*Drosophila funebris*). In this species no eye or body variations were induced in these experiments as in *D. melanogaster*, but only wings and bristles were affected, the relative stability of certain genes being apparently different in the different species. (After Timofeeff-Ressovsky, *Journal of Heredity*.)

very strong growing and dark green. Many crop plants are being subjected to X-rays in the endeavour to secure new and valuable variations with a greater rapidity.

Strangeways found, on subjecting his tissue cultures to X-rays, that the cells were least vulnerable when in a state of division. Once started, the divisions were apparently continued without interruption, but immediately before the division they proved to be exceedingly vulnerable and were much affected by the rays. This may be one reason why all cells are not equally affected when the embryos and other tissues are subjected to the rays, since presumably it would only be those cells which happened to be in this particular phase that would be affected (at least in the case of transmutations) and these would only form a small percentage of the whole. Work on the effect of external conditions on animal embryos shows that critical points occur in their development when they can be much affected, while in the intervals between nothing much occurs. These critical periods probably correspond to the transition periods between the three great phases of development, i.e. (1) stage of cell growth without differentiation, (2) the differentiation into definite organs, and (3) the appearance of function in the various organs.

There is, however, a much wider significance in this work on the effect of X-rays than that of a useful aid to geneticists and to plant and animal breeders to produce more frequent variations in their cultures. Recent work in physics has shown that various penetrating radiations known as cosmic rays occur throughout space, and some of these reaching earth may have had an effect on the evolution of living organisms by causing gene mutations and chromosome transmutations, since Millikan and Cameron found that about 11·4 atoms in every cubic centimetre of air are broken up by these rays every second. The probability that genes are compound molecules with a constitution of atoms, electrons and protons would make them more open to bombardment by these penetrating rays.

Apart from these external influences there is considerable scope for automutations in the genes themselves. When one contemplates the immense complexity involved in the reproduction division of a single gene with its included molecules, atoms, electrons, protons and perhaps neutrons, it would not be surprising if occasionally the normal equational division of the genes went wrong and a little more or a little less included in the new daughter gene. It is highly probable that certain mutations, e.g. multiple allelomorphs of a quantitative nature, may be due to an occasional inequality of division of the genes, especially if the theory of the step allelomorphs of the Russian geneticists be accepted. It is also conceivable that at times dividing genes may fail to separate, as in the observed non-disjunction of chromosomes, when a duplicated gene would arise giving a double or polyploid reaction.

The experimental application of X-rays and other radiations to the germ-cells of plants and animals has opened up a new field in biology of vast possibilities. By applying X-rays to the Fruit Fly (*Drosophila*), Muller succeeded in producing at will hundreds of gene mutations and chromosome transmutations of a similar nature to those which had occurred spontaneously in the genetical experiments of Morgan and others in normal conditions during many years. In other words, Muller artificially reproduced in a few minutes' treatment the mutations and transmutations that had taken seventeen years to accumulate in a natural way, besides other mutations new to science. Compared with the untreated controls, the use of X-rays increased the mutation rate about 150 times, representing an increase of about 15,000 per cent. The great majority of these mutations are of course quite useless, many being unviable or lethal; on the other hand, a few are really useful, viable and progressive and a distinct asset to the species. During the next ten years we may expect to see a speeding up of genetics by the use of X-rays, and it is possible that in the next twenty years more new and valuable mutations may be created by man than have arisen naturally in the whole of man's cultivations during the last 8000 years. Here at last man seems to have discovered the means whereby he can control evolution by proceeding to create new varieties, species and genera of plants and animals. Through long ages of secular time, millions of years before man himself appeared, these mutations and transmutations have been occurring in nature, and in accordance with the principles of natural selection the best fitted and adapted to the environment have

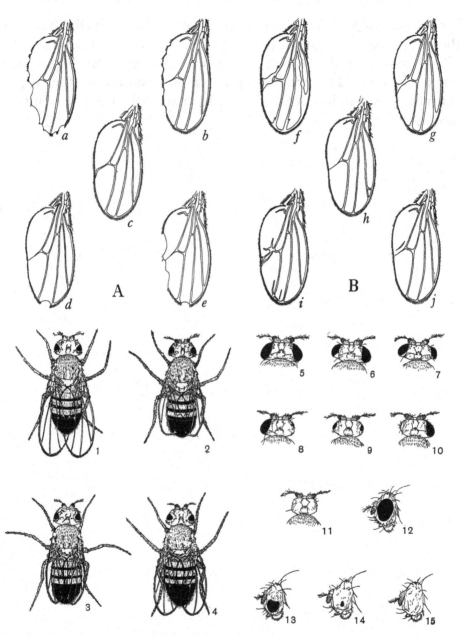

Fig. 146. Mutations in wing (venation and length) and eye characters induced in *Drosophila melanogaster* by radium irradiation. (After Hanson and Winkleman, *Journal of Heredity*.)

been preserved and the unfit and ill-adapted have been eliminated. In the future man will take a hand in the selection of his newly created species and natural selection will in course of time be displaced by human selection. The trend and progress of evolution will then depend entirely on the intelligence displayed by man in the exercise of his new powers.

CHAPTER XIII

GENES AND CHARACTERS

IN the map of the genes in the four chromosomes of the Fruit Fly
(*Drosophila melanogaster*) (fig. 30) it will be observed that each gene is
named in accordance with the particular character it represents in
the adult fly. From this it might be inferred by those unfamiliar with
genetical experiments that each gene determines and controls a single
unit-character in the organism. Such a conclusion would be far from
the truth, since experiments show in every case that has been fully investigated that more than one pair of genes are concerned with the
development of each character, and that their influence usually extends
to more than one character. Actually the whole complex of genes,
jointly and severally, in their interactions and reactions with one
another and with the internal and external environments, organise and
control the development and ultimately determine the characters of the
organism. It is true that a change in a gene or a pair of genes will alter
the balance of the complex so as to give rise to a difference in the end
result, and it is in this sense that the geneticist speaks of a gene determining a character. Thus, in conventional genetical shorthand, it is
commonly said that the wild agouti fur of the rabbit is determined by a
gene A, and in the sense that no rabbit can have agouti fur unless the
dominant gene A is present in the chromosomes of the cells, this is a true
statement. Such a simple account, however, is far from being the whole
story, and although it does not delude the geneticist it may mislead
those who are unfamiliar with the use of genetical formulae.

Genetical experiments show that, in the simplest case, at least four
pairs of genes are concerned in the organisation and development of the
wild agouti coat colour in rabbits and many other genes are also concerned.

(1) *Rabbits*

The familiar case of the agouti fur of the rabbit was one of the first
specific characters in animals to be genetically investigated (Hurst,

1904; Castle, 1905), and since then it has been exhaustively analysed by many experimenters in Europe and America. It may therefore be taken as an illustration of the genetical analysis of a specific character in an animal. Agouti fur is common to the rabbit, rat, mouse, guinea-pig and other rodents, and presents close genetical similarities in these different genera, indicating a common genetical basis and origin. In rabbits the colour of the hairs which go to make up the fur is closely associated genetically with the colour and pigmentation of the iris, much more so than in man, yet, as in other mammals, the rabbit presents in the genetics of its eye colours close and almost parallel analogies with those of man leading to a clearer knowledge of human genetics.

In appearance the agouti coat of the rabbit (fig. 147A) is a complex mixture of colour shades due to the variable deposition of black and yellow pigments in the hairs, distributed in the more or less regular pattern of a black base and apex with a yellow band between. An agouti hair may therefore be regarded either as a yellow hair based and tipped with black or a black hair banded with yellow; in either case it is a bicolor pattern rather than a single colour. Among the bicolor hairs that go to form the agouti coat there are a few long black hairs scattered along the back, loins, ears and forehead, which are known as "guard hairs", giving a wavy appearance to the coat known to fanciers as "ticking". Associated with the agouti coat are a grey-white belly and undertail, and brown duplex eyes so called owing to the presence of two distinct layers of pigment, one behind and one in front of the iris as in the brown duplex eyes of man (Hurst, 1908).

We have already noted that at least four pairs of genes are concerned in the agouti coat colour of the wild rabbit, and for convenience these genes may be called $AABBCCDD$. In a state of nature these four pairs of genes are all dominant and homozygous and consequently the individuals carrying them breed true to the agouti coat colour. Occasionally, however, these genes mutate to a recessive state both in nature and under domestication.

When the dominant gene A mutates to the recessive state a, and by inbreeding becomes aa, the genic constitution is changed to $aaBBCCDD$, and the wild agouti rabbit with light belly and undertail and brown duplex eyes is changed to a solid black rabbit with black brown eyes (fig. 148 b). When the dominant gene B mutates to b and becomes bb, the genic formula is $AAbbCCDD$, and the wild agouti is changed to a self-coloured golden yellow with yellow brown duplex eyes (fig. 149).

If, however, by any chance the genes A and B mutate simultaneously in a wild agouti, or separately in a self-yellow or a black rabbit, and become $aabb$, the genic formula will be $aabbCCDD$, and the agouti or the self-yellow or the black will each be changed into a tortoiseshell rabbit with yellow body and dark extremities (nose, feet, belly and under-tail), with black hairs in these parts and light brown duplex eyes. These four pigmented coat colours in the rabbit, agouti, black, self-yellow and tortoiseshell, are perfectly discrete characters easily distinguished at sight, although the two kinds of yellow, self-yellow and tortoiseshell, are often confused genetically, notwithstanding that their clear genic difference is that the self-yellow carries A and the tortoiseshell aa. The self-yellow may be either AA or Aa, since all these colour genes are completely dominant and a single dose of the dominant gene produces the same effect as a double dose, while the recessive gene is only effective with a double dose.

It will be observed that out of the six possible matings between these four pigmented coat colours, four are contrasting pairs of Mendelian characters, each pair with a genic difference, viz. agouti ($AABB$) and black ($aaBB$), agouti ($AABB$) and yellow ($AAbb$), black ($aaBB$) and tortoiseshell ($aabb$), and yellow ($AAbb$) and tortoiseshell ($aabb$). Each of these four matings gives Mendelian dominance in F_1, segregation of $3 D : 1 R$ in F_2 and purity in F_3 in the ratio of $1 DD : 2 DR : 1 RR$, in accordance with Mendel's laws. It will be noted that agouti and black differ in the same pair of genes (AA and aa) as yellow and tortoiseshell, but the two pairs produce different characters, because the former pair is BB while the latter pair is bb. In a similar way agouti and yellow differ in the same pair of genes (BB and bb) as black and tortoiseshell, but the two pairs produce different characters, because the former pair is AA while the latter pair is aa. From these results it is evident that in genetical research the finding of a single Mendelian difference is insufficient by itself and may be positively misleading, since the effect may be completely changed by the presence of other interacting pairs of genes.

The two remaining matings of the six, viz. agouti ($AABB$) with tortoiseshell ($aabb$) and black ($aaBB$) with yellow ($AAbb$), both present two and the same pairs of Mendelian differences $AAaa$ and $BBbb$ and both matings give identical results. The F_1 are all agouti ($AaBb$) in both matings, and in F_2 each mating gives a ratio of 9 agoutis : 3 blacks : 3 yellows : 1 tortoiseshell, as expected in cases of two Mendelian differences. The production of this ratio in experimental matings demonstrated the genetical constitution of these four pigmented coat colours in the rabbit (Hurst, 1907) and showed how each is produced by the interactions between the two pairs of genes AA and BB and their mutations aa and bb.

When, in a wild agouti of the constitution $AABBCCDD$, the dominant gene D mutates to the recessive state d, and by inbreeding becomes dd, the genic formula is $AABBCCdd$, and the agouti coat and brown duplex eyes are changed to a white coat and blue or grey simplex eyes as found in a White Beveren or White Viennese rabbit (fig. 147C). This mutation is distinct and important for two reasons: First, the white fur, which is due to lack of pigment in the coat hairs, shows that the development of pigment in the agouti is dependent on the dominant gene D, since in its recessive state dd no pigment is formed in the coat notwithstanding the presence of the dominant agouti colour genes $AABBCC$. Second, the blue or grey eye colour shows that some pigment is developed in the eyes, if not in the coat, and that in the recessive state dd the eyes are simplex while in the dominant state D the eyes are duplex. The eye colour in this case is of peculiar interest, since in both the rabbit and man the deposition of the purple-black posterior layer of pigment in the uvea and choroid behind the iris, with no trace of the brown anterior layer in front of the iris, constitutes a simplex eye which exhibits a shade of blue or grey according to the internal structure of the stroma of the iris. The brown duplex eye, on the other hand, is due to the presence of two layers of pigment, the posterior layer as in the simplex eye plus the anterior layer of brown pigment in front of the iris, making it duplex. The development of the brown duplex eye in both the rabbit and man is dependent on the *D* gene, and in the recessive state *dd* the eyes are blue or grey simplex in both (Hurst, 1908).

Finally, when, in a wild agouti of the constitution *AABBCCDD*, the dominant gene *C* mutates to the recessive state *c*, and by inbreeding becomes *cc*, the genic formula is *AABBccDD*, and the agouti coat and brown duplex eyes are changed to a pure white coat and pink eyes as found in complete albinos such as the Polish rabbit (fig. 147*B*). In these albinos no pigment is deposited in any part of the body, so that the coat is pure white and there is no anterior pigment in the front of the iris and no posterior pigment in the uvea and choroid of the eye, the transparent iris exhibiting the reddish pink colour of the blood-vessels. These results show that the development of the posterior layer of pigment in the uvea and choroid of the eye, and consequently the blue and grey simplex eye of the rabbit and man, is dependent on the dominant gene *C*, since in its recessive state *cc* no pigment at all is developed in the eyes notwithstanding the presence of the dominant agouti colour genes *AABB* and the duplex eye colour genes *DD*. Experiments in crossing White Beverens and albino Polish with the four pigmented forms and with one another show that the white breeds, although unable to form pigment in their coats (fig. 147), are both able to carry the coat colour genes *Aa* and *Bb* in any combinations, but these can only come into action in the presence of both *C* and *D*. It is only when the genes *C* and *D* are present together that rabbits have coloured coats and brown duplex eyes, and it is only when the genes *AA* and *BB* are also present with *CC* and *DD* in a homozygous condition that we get the true-breeding wild agouti rabbit as it exists in a state of nature.

Some interesting results were witnessed in these experiments. For instance, black crossed with white (often quoted as an illustration of simple Mendelism) gave most variable results although all followed strictly the Mendelian system. Thus, when a pure black of the genic constitution *aaBBCCDD* was mated with an albino of the constitution *aaBBccDD*, the F_1 offspring were black and F_2 gave 3 blacks to 1 white in accordance with the simple Mendelian rule. But when the same black was mated with an albino of the constitution *AAbbccdd*, the F_1 offspring were all agouti instead of black and F_2 produced agoutis, blacks, yellows, tortoiseshells, White Beverens and albinos instead of blacks and whites. In the same way a pure agouti *AABBCCDD* mated with an albino *AABBccDD* gave agoutis in F_1 and 3 agoutis to 1 albino in F_2, but mated with an albino *aabbccdd* it gave agoutis in F_1 and all the six colours in F_2. Similarly a pure yellow *AAbbCCDD* mated with an albino *AAbbccDD* gave yellows in F_1 and 3 yellows to 1 albino in F_2, but mated with an albino *aaBBccdd* produced agoutis in F_1 and all the six colours in F_2. A pure tortoiseshell *aabbCCDD* mated with an albino *aabbccDD* gave tortoiseshell F_1 and 3 tortoiseshells to 1 albino in F_2, but mated with an albino *AABBccdd* it produced agoutis in F_1 and all six colours in F_2.

The same tortoiseshell mated with an albino *AAbbccDD* gave a yellow F_1, with yellows, tortoiseshells and albinos in F_2, but mated with an albino *aaBBccDD* the F_1 was black and F_2 gave blacks, tortoiseshells and albinos. The above matings were made in a series of tests with Polish albinos at Burbage, together with many others, but owing to the War the results were never published. Altogether eight genic types of albinos were found, and perhaps the most striking results were obtained by crossing the White Beverens with the Polish albinos. Both bred true as separate breeds, the Beverens with pure white coats and blue or grey simplex eyes and the Polish with pure white coats and pink eyes. A White Beveren of the constitution *aaBBCCdd* mated with a Polish albino of the constitution *aaBBccdd* gave White Beverens in F_1 and 3 White Beverens to 1 Polish albino in F_2, but the same White Beveren mated with a Polish albino of constitution *AAbbccDD* gave agouti selfs in F_1 (fig. 147) and all the six colours, agouti, black, yellow, tortoiseshell, White Beveren and White Polish, in F_2.

These results fully demonstrate that at least four pairs of genes *AABBCCDD* are concerned in the development of the agouti coat in rabbits.

Many other pairs of genes are known to be associated with coat colour; one pair, which may be called *FF*, changes agouti into cinnamon, and black into chocolate brown, when in the recessive state *ff*.

Another, *GG*, in the recessive state *gg*, reduces the amount of pigment deposited in the hairs, leading to a dilution of the normal dense colour represented by *GG*. Thus a mutation of the wild *G* to *g*, becoming *gg* by inbreeding, changes agouti to blue-grey, black to blue, chocolate to "lilac", yellow to creamy fawn, and tortoiseshell to blue-fawn. The

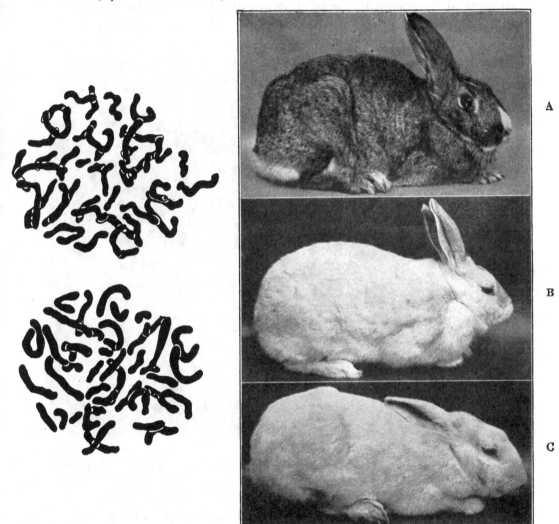

Fig. 147. *A*, Agouti Rabbit from the crossing of *B* and *C*, both of which are pure breeding white rabbits with pink nulliplex and blue simplex eyes respectively, carrying genes nevertheless which, when brought together, produce agouti coat and brown duplex eyes. At the side two sets of somatic chromosomes of the rabbit (44). (After Castle and Painter.)

workings of this pair of genes *gg* provide an interesting example of the multifarious effects in a single direction of a pair of recessive genes in its

interactions with various dominant colour genes. Other recessive genes
have given rise to colour mutations, some of which appear to be unilocal
multiple allelomorphs to the wild agouti and its mutations, e.g. Chin-
chilla which is agouti minus the yellow with the black full or reduced,

a　　　　　　*b*　　　　　　*c*　　　　　　*d*

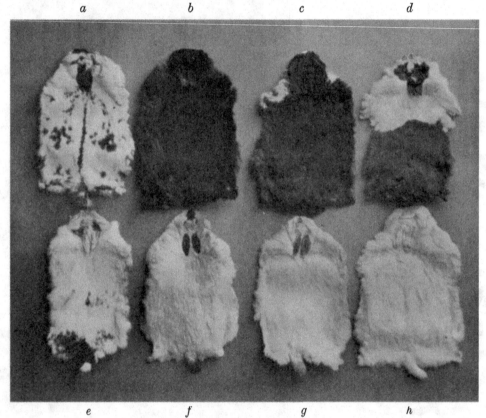

e　　　　　　*f*　　　　　　*g*　　　　　　*h*

Fig. 148. Coat patterns in Rabbits. *a*, English-marked; *b*, Self-coloured (Grade 10); *c*, Deep
Dutch (Grade 8); *d*, Dutch-marked (Grade 5); *e*, Light-spotted Dutch (Grade 2); *f*, Himalayan; *g*, F_1 Himalayan × Clear White; *h*, Clear White Albino (Grade o).

Himalayan white with red eyes and coloured extremities (fig. 148*f*),
making a series of six unilocal allelomorphs of the gene *C*; Japanese,
with mosaic colour pattern; rufous modifiers in the Belgian Hare
agouti, and New Zealand Reds. In the Belgian Hare rabbit there
appear to be three rufous modifiers of the wild agouti which are recessive in F_1, exhibit the Engledow "shift" in F_2 and appear in full

strength in F_3 and F_4 by the cumulative effect of the three genes; two of these rufous modifiers have been found in New Zealand Reds, which otherwise is a self-coloured yellow of the constitution *AAbbCCDD*.

There are also at least six pairs of genes which are "dominant" or epistatic to agouti, including three pairs for "silvering", which behave like grey and roan coat colours in horses; one pair for coloured and tan pattern; one pair for "dominant black", which changes agouti to steel-grey (fig. 151); and one (possibly more) pair for the English spotted pattern (fig. 148 a).

The coloured and white pattern mutations in the rabbit present a remarkable series ranging from the "dominant" white English with white fur plus a little colour, through the self-coloured wild type with no white markings to the "recessive" white Dutch with white fur plus a little colour. Although this series of minus, plus and minus coloured areas is completely continuous except for a rather considerable gap between the most heavily spotted English and the wild self-coloured, yet the English and Dutch patterns are peculiarly distinct and discontinuous in character both from one another and from the wild self-coloured. Fundamentally the "dominant" English is a white rabbit with a peculiar pattern of spots arranged in chains, a butterfly marked nose and dark ears and eyes in some respects not unlike a lightly marked Dalmatian dog, while the "recessive" Dutch is a coloured rabbit marked with white in a definite pattern with a blaze and the whole of the fore parts white except the ears, cheeks and eyes, and the whole of the hind parts coloured except the feet or "stops". The same Dutch pattern is found in many mammals, including mice, guinea-pigs, dogs (e.g. fox-hounds and terriers), Hereford cattle, and to a less degree cats and horses with white blaze and socks. The exhibition English (fig. 148 a) is a white rabbit rather lightly marked with coloured spots and is always heterozygous for self-coloured pattern. Castle, who first raised the homozygous form, found it to be very lightly marked, closely approaching the lightest form known as white English. The exhibition Dutch (fig. 149) is about one-half coloured and one-half white, the white area being chiefly in front; it is usually heterozygous, but can be bred fairly true to type. Thanks to the patient experiments and ex-

Fig. 149. Ch. Burbage Yellow Premier. The first Champion Yellow Dutch (Grade 5).

Fig. 150. Light-spotted Dutch of Grade 2.

Fig. 151. Ch. Mikado. A steel-grey homozygous Standard Dutch of Grade 5.

Fig. 152. Blackburn Lassie. A black heterozygous Mock Dutch of Grade 5.

haustive analysis of the Dutch rabbit by Punnett and Pease (1925), we are able to add four more pairs of genes to the genic constitution of the wild self-coloured rabbit, distinguished by them as *NNPPSSTT*. These genes only come into action when associated with the colour genes *AABBCCDD* or their mutations. There is, however, an important difference in the action of these two sets of gene complexes, in so far that while the colour genes are completely dominant, so that *AaBbCcDd* produce the same result as *AABBCCDD*, the Dutch pattern genes are incompletely dominant, so that *NnPpSsTt* gives a different result from *NNPPSSTT*. In the former case a single dose of the dominant gives the same effect as a double dose, while in the latter case the double dose gives about twice the effect of a single dose. On the whole the pattern genes have a multiple or cumulative effect on the extent of the coloured areas, so that when all are present in a dominant state *NNPPSSTT* the rabbit is self-coloured with no white markings, while in a recessive state *nnppsstt* the rabbit is almost white with a few coloured spots as in the white Dutch which is the lowest colour grade. The Dutch pattern genes, however, differ from the ordinary multiple genes found, e.g. in Red Wheats, in so far that their effect, though cumulative and in the same direction of more pigment, is apparently unequal in value in every case. Thus the gene *S* has considerably more effect than the gene *T*, though both make for more colour and less white and *SS* and *TT* have a greater effect than *Ss* and *Tt*. *PP* has apparently more effect than *SS* and *TT* combined, and *Pp* has less influence than *PP*. *NN* is less effective than *PP*, though apparently necessary to the development of the all-coloured self. Consequently the various interactions between these four pairs of genes produce highly complicated results, though all follow strictly the Mendelian system.

Seeing that the different patterns of Dutch rabbits depend largely on the relative amounts of colour and white in the coat, it has been found convenient to classify the continuously variable Dutch phenotypes into 10 colour grades (Hurst, 1913 *b*).

Grade 10 are All-coloured selfs of the wild type with no trace of white Dutch markings (fig. 148 *b*), and their genic constitution is *NNPPSSTT* plus the colour complex *AABBCCDD* or its mutations which are common to all the 10 grades. This type of self-colour is apparently common to the wild agouti Flemish Giant, Belgian Hare, Coloured and Tan, Himalayan, Angora and probably the English selfs. Other kinds of selfs, e.g. the Blue Beveren, are indicated of the genic formula *HHPPSSTT*, where *NN* is displaced by *HH* and there may be others of different constitutions.

Grade 9 are Marked selfs, almost self-coloured but lightly marked or variably tipped with white on the nose, breast, forehead or front paw. To this grade belongs the rare St Nicholas breed which was derived from the Blue Beveren self by the mutation of the *P* gene to *p*, becoming *pp* by inbreeding.

Grade 8 are Deep Dutch, the white markings extending to the legs, shoulders, breast and neck, often showing a narrow blaze (fig. 148 *c*).

Grade 7 are Dark Dutch, which have more white than Deep Dutch and less white than the Typical Dutch.

Grade 6 are Mismarked Dutch with patches of misplaced colour, otherwise they are Typical Dutch.

Grade 5 are Typical Dutch with equal proportions of coloured and white areas but more regularly marked than either grade 6 or 4 (figs. 148 *d*, 149, 151, 152).

Grade 4 are White Marked Dutch with small bands or areas of white in the coloured hind parts, otherwise they are Typical Dutch.

Grade 3 are Dark Spotted Dutch with more white than colour, the latter being broken up into irregular spots and blotches.

Grade 2 are Light Spotted Dutch with more white and fewer and smaller coloured spots (fig. 150).

Grade 1 are White Dutch, the lowest coloured form, in which the coat is pure white except that odd spots of colour are found on the rump or round the eye, which is heterochromic and duplex, often blue or grey but never quite simplex.

Grade 0 are clear white albinos with pink eyes (Polish), which appear occasionally in pedigree Dutch stock following a mutation of the *C* gene (figs. 147 *B*, 148 *h*).

Table 1 shows the various genotypes and phenotypes so far found in the 10 grades of the Dutch rabbit and also the different grades these produce when inbred with their own genotype. It will be noted that grade 1 is the only grade which always breeds true, but all the grades have genotypes which breed true or within the limits of the nearest grade, and grade 7 is the only grade which throws all the 9 grades.

TABLE 1. *Genes and characters in the Dutch rabbit*

Grades	Genotypes	Phenotypes	Grades produced when inbred
10	NNPPSSTT	Wild self-coloured	10
10	HHPPSSTT	Beveren self	10
10	NnPPSSTT	Self-coloured	10, 9
9	NnPpSSTT	Marked self	10, 9, 8, 7, 6, 5, 4
9	nnPPSSTT	Marked self	9
9	nnPPSStt	,,	9
9	nnPPSsTt	,,	9, 8
9	HHppSSTT	St Nicholas	9
8	nnPPSstt	Deep Dutch	9, 8
8	nnPPssTT	,,	8
8	nnPPssTt	,,	8
8	nnPPsstt	,,	8
8	NNppSSTT	Castle's "Tan" Dutch	8, 7
7	nnPpSSTT	Dark Dutch	9, 7, 6, 5, 4
7	nnPpSSTt	,,	9, 7, 6, 5, 4, 3
7	nnPpSStt	,,	9, 7, 3
7	nnPpSsTT	,,	9, 8, 7, 6, 5, 4, 3, 2,
7	nnPpSsTt	,,	9, 8, 7, 6, 5, 4, 3, 2, 1
7	nnPpSstt	,,	9, 8, 7, 5, 3, 2, 1
7	NNppSSTT	Castle's "Tan" Dutch	8, 7
6	nnppSSTT	Mismarked Dutch	6, 5
5	nnppSSTT	Typical Dutch	6, 5, 4
5	nnPpssTT	Mock Dutch	8, 5, 2
5	nnPpssTt	,,	8, 5, 2, 1
5	nnPpsstt	,,	8, 5, 1
4	nnppSSTT	White Marked Dutch	5, 4
3	nnppSSTt	Dark Spotted Dutch	6, 5, 4, 3, 2
3	nnppSStt	,, ,,	3
3	nnppSsTT	,, ,,	6, 5, 4, 3, 2
2	nnppSsTt	Light Spotted Dutch	6, 5, 4, 3, 2, 1
2	nnppSstt	,, ,,	3, 2, 1
2	nnppssTT	,, ,,	2
2	nnppssTt	,, ,,	2, 1
1	nnppsstt	White Dutch	1

Grade 5, with typical Dutch markings, consist of two distinct classes, the standard genotype which breeds true to the narrow limits of 4 and 6 (fig. 151) and the Mock Dutch of three genotypes that throw grades 8, 5, 2 and 1 (fig. 152). In outward appearance these four genotypes are indistinguishable, although, owing to the presence of the gene P in the Mock Dutch, no specked eyes are found in them, while in the standard genotype about 70 per cent. have *heterochromia iridis*. Out of nine champion and prize Dutch rabbits tested at Burbage before the War, six proved to be Mock Dutch, while three were of the true-breeding standard genotype (Hurst, 1925).

Since these specimens were collected from different areas in Great Britain, in order to represent as many strains and colours as possible, it is clear that a large proportion of strains of exhibition Dutch are of the Mock Dutch type. One explanation of this may be that for many years Dutch breeders have been trying to breed out of their strains the specked eye, which is regarded by most judges as a serious fault; it is due to local defects of the anterior layer of brown pigment on the outside of the iris causing the purple-black posterior layer of pigment in the uvea and chorion to show through in the form of blue specks or larger areas. This is a common defect in the true-breeding Dutch genotype, which is pp, but has not been found in the Mock Dutch, which is Pp. It may be that in trying to eradicate this defect breeders have unconsciously selected animals of the Mock Dutch Pp type, and in this way the true-breeding genotype pp is being gradually eliminated in exhibition strains.

The Self-coloured grade 10 (fig. 148 b) usually breeds true but occasionally, when heterozygous for n, s or t, may throw grade 9 lightly tipped with white.

The rare St Nicholas rabbit of grade 9 was derived from the Blue Beveren self, which apparently differs from the ordinary self by the substitution of HH for NN.

Table 1 shows how the various grades may have originated from the wild self or from one another by the mutations of the four pairs of genes to a recessive state, and finally shows that the wild self-coloured coat is established and breeds true only when the four pairs of genes $NNPPSSTT$ are present in a dominant and homozygous state.

In addition to the ordinary colour and pattern genes there are other genes which affect the agouti colour by changing the length, texture and quality of the fur peculiar to the rabbit. The most important of these are the two pairs LL and RR, which in a recessive state ll and rr produce great changes in the fur, making a much longer or much shorter coat than in the wild species and also changing the shade of the coat colour. ll produces the long hair of woolly texture of the Angora rabbit, while *rr* produces the very short mole-like hair of the Rex rabbit. Both these extremes of long and short hair are recessive to the normal wild fur, which is of the genic constitution *LLRR*. Altogether there are at least twenty-six pairs of genes known to be concerned with the fur of the rabbit. Deducting the six "dominant" or epistatic mutations and the eight "recessive" mutations which are not yet completely understood, some of which appear to be unilocal allelomorphs, leaves us with at least twelve pairs of genes which are definitely concerned with the wild agouti fur of the rabbit. Since each recessive gene originally arose either directly or indirectly from a dominant wild gene, it follows that the agouti fur must be genetically composed of at least twelve pairs of corresponding dominant genes interacting in unison with one another to produce the agouti coat. This specific complex of genes is preserved in nature by the natural selection of the individuals carrying them in a dominant and homozygous state. Occasionally in a wild state, and more frequently under domestication and more often still in genetical experiments, the wild homozygous specific complex is broken down by varietal mutations which follow Mendel's laws of heredity, and by human selection, preservation and breeding are established as domesticated races, and new combinations of these mutations are being constantly created by cross-breeding and preserved by inbreeding.

So far as these twelve pairs of genes are concerned, the genic formula of the wild agouti fur in the rabbit may be written as follows:

$$\{[(AA + BB + CC + DD) + (FF + GG)] \\ + [(NN + PP + SS + TT)] + (LL + RR)\},$$

where A–G are colour genes, N–T pattern genes of colour and white, and L and R genes for length and texture of fur. No attempt has been made to represent linkage of genes in the same chromosome, since exhaustive investigations have not yet been made.

So far as the linkages are known, these twelve pairs of genes are carried in ten of the twenty-two pairs of chromosomes found in the rabbit.

These genetical experiments with rabbits give us an insight into the nature of specific characters in general and how they work in the animal under the control of natural selection. It is evident that intrinsically there is no difference in kind between varietal and specific characters, since both are organised and determined by different complexes of genes. They do differ, however, in degree, since specific characters are usually represented by larger complexes of dominant homozygous genes, while varietal characters are represented by smaller complexes of heterozygous and recessive genes, representing for the most part recessive mutations of the dominant specific genes. When a dominant specific gene mutates it becomes heterozygous and automatically changes into a varietal gene, and when several specific genes mutate a varietal complex of genes is established by inbreeding. In this way specific characters may be said to represent the *status quo* under natural selection, while varietal characters represent potentialities of future species under the changing conditions of life and natural selection.

To take a simple and purely hypothetical case. A colony of wild agouti rabbits occupies a normal area of mixed ground where burrowing is possible, in which their agouti coats fit into the landscape colour scheme so that the majority of individuals are adapted to survive and to increase their numbers under the eliminating conditions of natural selection. On the north of this area are yellow sand dunes of such a nature that a yellow rabbit would be more adapted to survive and increase than an agouti. The agouti rabbits increase and extend over both areas, dropping here and there in the burrows yellow varietal mutants of the genic constitution *AAbbCCDD* instead of the agouti constitution *AABBCCDD*. In course of time, if natural selection is duly effective, the yellow rabbits will increase more on the yellow sand dunes than on the mixed ground, where the agouti individuals will maintain their position and numbers. As the number of yellow individuals increases on the sand dunes the yellow matings will become more frequent, until in course of time a yellow race will be established on the sand dunes. If by any chance the two areas become isolated from one another by an intrusion of the sea or some other means, a yellow race of rabbits will become definitely established and fixed in this habitat.

Although the above case of the rabbit is purely hypothetical, it is closely analogous and almost parallel with the actual observations and experiments of Sumner with the Florida Mouse (*Peromyscus polionotus*) (Sumner, 1930). On Santa Rosa Island in the Gulf of Mexico, which is a narrow bank of white sand stretching along the coast for about 50 miles and separated from the mainland by a narrow strait, there is a yellowish white race of this mouse distinguished from the dark-coloured typical race on the mainland by the name *leucocephalus*. The dark typical form of the species is found on dark soil at some distance from the sea and an intermediate form is found between them near to the sea, and there can be little doubt that the light-coloured race on Santa Rosa Island is an adaptation to its surroundings, which has become established owing to its isolation from the dark typical form of the mainland. Genetical experiments show that the differences in colour between the two races are dependent on Mendelian genes and that the light-coloured island race is a mutation from the dark typical form on the mainland. In such cases the mechanism of evolution is the varietal mutation of the specific gene complex. Under changed conditions of life these mutants are better adapted to survive and increase than the original species adapted to the old conditions and ultimately may with other mutations and transmutations evolve into a new adapted species with a new specific complex of dominant homozygous genes. The evolution of species thus becomes a series of rhythmic cycles, species—mutational varieties—subspecies—species, under the control and direction of natural selection through the mechanism of the genes and chromosomes.

Other rodents besides rabbits, such as mice, guinea-pigs and rats, have analogous sets of gene complexes organising and controlling their specific fur, though naturally the sets are not parallel, their differences in this respect representing a precise measure of their generic types of coat coverings. Analogous examples of large but limited gene complexes organising the specific characters in other genera, orders and classes of animals might be given, especially in birds and insects; indeed it may be said that such cases are found whenever a species has been genetically analysed on a large scale.

Owing to technical facilities species of plants have been worked out genetically on a much larger scale than animals generally, though a brilliant exception must be made in the case of the Fruit Fly (*Drosophila*), which for many years has led the way in all advanced genetical research owing to the rapidity of its generations. Plants, however, have been bred on a wider scale in many different genera, families and classes, and similar large but limited gene complexes controlling the specific characters have been found which are in every way analogous to those found in animals.

(2) *Sweet Peas*

Among plants, one of the earliest and best known cases is that of the Sweet Pea (Bateson and Punnett, 1905; Hurst, 1910), and the chief specific characters that have been worked out are the habit of growth, and the structure, form and colour of the flowers.

These experiments show that the species complex of the Wild Sweet Pea (*Lathyrus odoratus*) is composed of at least ten pairs of interacting genes located in seven pairs of chromosomes. In the wild type all the ten pairs of genes are homozygous dominants, breeding true to the specific complex, except for occasional mutations. In each case the dominance is complete, so that one dose of a dominant Tt gives the same result as two doses TT. The genic formula of the wild type may be written:

$$\{(TT + PP) + KK + XX$$
$$+ [WW + (CC + RR) + (BB + DD + LL)]\},$$

where T and P organise the habit of growth, K the structure of the keel of the flower, X the shape of the pollen grains, and W, C, R, B, D and L the colour of the flowers. T determines tall stems, several feet high, with elongated nodes, while the recessive mutation tt produces a dwarf variety of the species, a few inches high, with short stems and nodes.

The P gene determines a prostrate habit of growth, causing the stems to lie flat on the ground. In the wild species and the tall sweet pea of gardens, which are $TTPP$, the long stem runs along the ground until it finds a support to which to cling and climb. $ttPP$ produces the cupid sweet pea of gardens, a dwarf prostrate variety a few inches high, which is unknown in a wild state and which appeared first in an American garden. The pp mutation organises an erect habit of growth, so that $TTpp$ produces the bush sweet pea, which is tall, erect and bushy but not climbing. The double recessive *ttpp* produces the dwarf erect cupid which differs from the ordinary cupid in its being erect instead of prostrate. It is evident that the bracketed ($TT + PP$) genes and their recessive mates (*ttpp*) work together in close association, notwithstanding that they appear to be located in different pairs of chromosomes.

The K gene organises the structure of the keel of the flower, so that the male and female organs are tightly clamped in the apices of the keel petals, ensuring complete self-fertilisation. In the wild species the

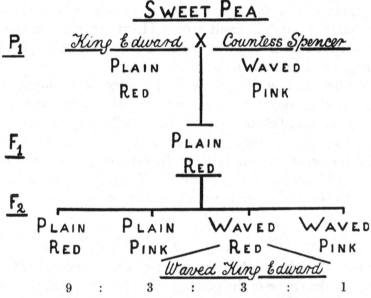

Fig. 153. Showing the origin of the "Waved King Edward" Sweet Pea by recombination of characters following Mendelian segregation.

clamped keel is associated with plain, unwaved standard and wing petals, the standard flat and erect with foreshortened apex, which is often notched. In gardens, however, the clamped keel is now more often associated with a hooded standard, which is itself a mutation of the wild flat erect standard. The mutation *kk* produces a more or less open keel, in which the male and female organs are only loosely associated in the keel petals and the pistil often protrudes, leading to cross-fertilisation. The open keel is associated with a waving or frilling of the standard and wings, formerly known in gardens as the "Spencer" sweet pea but now regarded as the standard of all exhibition sweet peas. This mutation is unknown in a wild state and first appeared in Lord Spencer's garden at Althorp Park in the pink variety known as Countess Spencer. At that time Mendelian experiments with sweet peas were under way a few miles away, and in a few years by their aid this mutation kk was added to all the best of the garden varieties (fig. 153), and it may be claimed that all the beautiful modern varieties of sweet peas have been built up in a few years by the application of Mendel's laws, and, what is perhaps more important, the seed stocks of the wholesale growers in Europe and America have in a few years been purified by the adoption of genetical methods (Hurst, 1913, 1925).

The gene X determines the shape of the pollen grains, so that they are long and usually 3-pored. This is the normal wild form. The mutation xx produces round-shaped pollen grains which are usually 2-pored.

The colour gene W organises the white or colourless plastids of the petal cells of the flower, forming a white ground colour for the red and blue sap-colours as in the wild sweet pea, which is purple and blue on a white ground colour. The mutation ww produces yellow plastids in the petal cells, forming a cream ground colour for the red and blue sap-colours in many garden varieties and an ivory-cream flower when the sap-colours are not developed. This cream mutant is not known in a wild state and appeared in gardens about the middle of the nineteenth century. This gene for plastid colour interacts with the five sap-colour genes to produce the complete flower colour, though otherwise it seems to be quite independent of them.

The remaining five pairs of colour genes $CRBDL$, which together organise the production and distribution of the sap-colours in the flower, differ from the other genes, inasmuch as they are remarkably inter-dependent in their relations and reactions to one another.

Thus C and R are only active in the presence of one another, and when they come together they produce red sap-colour in the flower.

B is only active in the presence of both C and R, and when the three genes come together they produce purple sap-colour in the flower.

D is only active in the presence of C and R, when they produce an intensification and deepening of the colour.

L is only active in the presence of *C*, *R* and *D*, when they produce bicolors with dark standards and light-coloured wings as in the wild type.

On the other hand *cc* and *rr* in any combination, e.g. *ccRRBBDDLL* or *CCrrBBDDLL*, produce no sap-colour but all white flowers with *W* and all cream flowers with *ww*. *C* and *R* with *bb* in any combination produce red flowers and with *dd* the flowers are white tinged or flushed with red, and with *B* added they are purple picotees. *ll* with *C*, *R* and *D* produce red selfs, with *B* added they are purple selfs, and so on. The

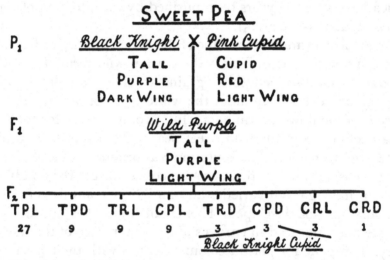

Fig. 154. The origin of the "Black Knight Cupid" Sweet Pea by recombination of characters after segregation. *T* = tall; *C* = cupid; *P* = purple; *R* = red; *L* = light wing; *D* = dark wing.

whole of these reactions of the colour genes, although apparently very complex, are in reality quite simple and orderly, following the Mendelian system throughout.

Finally, the chocolate purple standard and purple blue wings of the wild sweet pea are only produced when the five dominant sap-colour genes *CRBDL* are present in association with the plastid colour gene *W*, and in the wild species these genes are all homozygous, *CCRRBBDDLLWW*, and breed true as a specific character.

These genetical experiments with sweet peas and rabbits are of fundamental importance, since they demonstrate that characters are organised, determined and controlled by complexes of genes rather than by pairs of genes. The end result is not due alone to the independent action of pairs of genes, nor to the mere sum of their separate actions added together, but to these important factors must be added the effect produced by the definite reactions and interactions of these pairs of genes one with another. Recent work has further emphasised the vital importance of the combined effect of the whole complex of genes on the development of a single character. The end product is in fact a balance of the whole complex of genes interacting with one another and with the internal and external environments. Consequently, each individual, form, variety, species, genus and higher group in plants and animals has its own gene complex, whose peculiar genic balance determines its characters.

In our genic analyses we are apt to regard $A + B$ as merely the sum of A and B, when in reality we find that $A + B = X$; this result, though analysable into A and B, is something more, and in our calculations we have omitted to evaluate the interaction involved in the plus sign. Indeed, the progress of Creative Evolution depends to a large extent on the formula $A + B = X$. The consequences of these findings are far-reaching in their effects on genetical studies of plants and animals and are of paramount importance to studies in the field of human genetics.

(3) Man

Owing to the extensive experimental facilities available in the studies of plant and animal genetics which are not possible in the study of human genetics, the advances in the former have naturally quite outstripped the latter. Unless a serious attempt is made soon to bring investigations up towards the level already attained in plant and animal genetics, future eugenic efforts to promote the welfare of the race will be abortive. It is necessary to realise that at the present time research in human genetics, especially in the direction of preventive medicine, is hardly more advanced than it was a quarter of a century ago, when genetics was in the infant stage of dealing with single Mendelian differences. On the medical and statistical side admirable studies have been made of hundreds of family pedigrees, showing the incidence of a large number of diseases and defects which have been interpreted as simple dominant and recessive characters due to Mendelian differences.

These simple interpretations have been widely accepted by pathologists, eugenists and social workers, and in some cases practical social and legislative action has been taken on the strength of these interpretations. To the geneticist of experience such action seems to be unduly precipitate, the consequences of which, if carried out on an extensive scale, may prove to be disgenic rather than eugenic in many cases. It is necessary therefore to emphasise the importance of recent advances in plant and animal genetics and to apply them so far as it is possible to human genetics.

One of the early discoveries of Mendelian characters in man was that of duplex and simplex eye colour (Hurst, 1907). In a popular, though unfortunately inaccurate, way of speaking, brown eyes are said to be dominant and blue eyes recessive. The difficulty is that though all brown eyes are duplex, all blue eyes are not simplex. The Mendelian difference consists in the presence of two layers of pigment in duplex eyes and only one in simplex eyes. In duplex eyes there is a posterior layer of purple-black pigment in the uvea and choroid behind the iris and an anterior layer of brown pigment in front of the iris, while in blue or grey simplex eyes there is only one layer the posterior pigment, the anterior brown layer being absent. Many so called "blue" and "grey" eyes, however, are not simplex but low grade duplex eyes with traces of anterior pigment, and these can easily be distinguished with a strong hand lens.

In the experiments with rabbits noted above we have seen that the genes representing eye colour in rabbits and man are not only analogous but in many respects are strictly parallel, so that as far as the basic characters of eye colour are concerned the same letters representing the corresponding genes can be used to cover the similar data in rabbits and man. Thus the gene D may be said to organise the production of the brown layer of pigment on the front of the iris, while the gene C organises the purple-black layer of pigment behind the iris. The gene A determines the light shade of yellow-brown pigment on the front of the iris but only when C, D and b are present. The gene B determines the dark shade of black-brown in the same layer but only in the presence of *C*, *D* and *a*. Consequently, the genotype *CCDDAABB* produces an ordinary brown duplex eye, while *CCddAABB* produces a blue or grey simplex eye and *ccDDAABB* gives an albino nulliplex or pink eye with no trace of pigment. The genotype *CCDDAAbb* produces a yellow-brown duplex eye while *CCDDaaBB* gives a dark black-brown duplex eye, and so on through the various genotypes of these phenotypes. The basic genic formula for eye colour in rabbits and man can therefore be written [*CC* + *DD* + (*AA* + *BB*)] where *C* represents simplex, *D* duplex, *A* yellow-brown and *B* black-brown eye colours.

From the evidence in the rabbit and other organisms investigated we may reasonably infer that this basic formula for eye colour with its homozygous dominant genes represents the primitive wild type in man. This type of ordinary brown duplex eyes is common to many races of man with white, red, yellow or brown skins, but it is not usually found in the full-blooded negro, who is distinguished by dark black-brown eyes of constitution *CCDDaaBB*. From this it would appear that the negro is not the primitive type of man as some anthropologists have suggested. So far as the basic characters of eye colour are concerned it is clear that the genes are parallel in rabbits and man but in other respects there are distinct differences. Thus in the rabbit the basic eye colour genes are usually closely associated with corresponding coat colour genes, e.g. simplex eye colour with albino hairs, while in man simplex eye colour is associated with pigmented hairs or partial albinism and the genes for hair colour are often independent of those for the basic eye colours. In the variable shades and grades of eye colours in both man and the rabbit other modifying colour and pattern genes are clearly concerned, and from the published data in man (Davenport, 1907; Hurst, 1907, 1908, 1912) (fig. 155), there can be little doubt that these grades are dependent on a small number of modifying genes interacting with the four basic genes.

Three distinct systems of the working of modifying genes have so far been recognised in genetical experiments.

First, in the original case of the discovery of multiple genes by Nihlsson Ehle in which the three pairs of genes *R*, *S* and *T* that produce

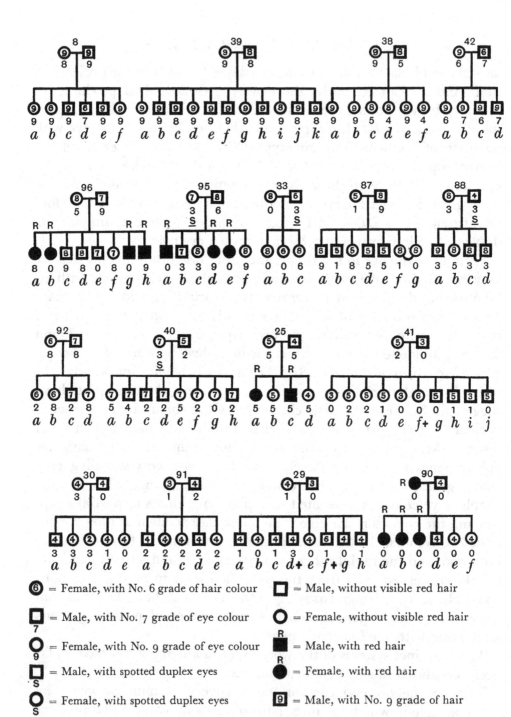

Fig. 155. Examples of eye colour and hair colour in Leicestershire families.

red grain in wheat have an equal cumulative effect, so that *RrSsTt* produces a medium shade of red, *RRSSTT* the darkest shade of red and *Rrsstt*, *rrSstt* and *rrssTt* the lightest shade of red, while *rrsstt* is white. In this way the three pairs of genes produce a graded series of shades from white to deep red, the frequencies of which, 1, 6, 15, 20, 15, 6, 1, fit the curve of probability. For many years geneticists have utilised Nihlsson Ehle's polymeric scheme of equally cumulative dominant factors to cover the genetics of graded and measurable characters of size, weight and speed which present a graded series of results. This scheme, however, has been applied chiefly to grades of dominant characters.

Second, the brilliant genetical analysis of the Dutch rabbit by Punnett and Pease (1925) provides a scheme for the analysis of modifying genes in recessive characters. We have already seen that this scheme involves a series of dominant genes which produce an unequal cumulative effect in the same direction, giving a complicated and over-lapping series of phenotypes, but the genotypes are the same and follow the usual Mendelian system.

Third, a similar scheme for the analysis of modifying genes in re-cessive characters has been employed with signal success by Philip-tschenko (1927), the Russian geneticist, in his remarkable analysis of the genetics of the famous "Marquis" wheat. This system differs from that of Punnett and Pease in so far that it involves an equal cumulative effect in the same direction of dominant (and recessive) genes modifying a recessive character. Philiptschenko rounds off his research by applying the results to an interpretation of the genetics of musical ability in man, which so far fits the known data and is certainly the most stimulating and encouraging work in human genetics that we have had for many years.

This important paper was originally published in Russian and came into the hands of Prof. Engledow of Cambridge, and, thanks to his in-terest and that of Prof. Sir Rowland Biffen of the Imperial Bureau of Genetics, Cambridge, an excellent English translation has been made by Dr Hudson, the Assistant Director of the Bureau.

After several years of experimental crossings and inter-breedings among the different sub-species, varieties and pure lines of the Bread Wheats (*Triticum vulgare*), Philiptschenko succeeded in identifying six pairs of genes which in combination produce the peculiar broad-shaped grains and glumes of the famous "Marquis" wheat. This super-wheat originated in Canada from a single grain selected by C. Saunders in 1903 from a hybrid obtained by his brother, A. Saunders in 1892, and has proved to be of international value, being cultivated with success over vast regions in Europe, Asia and America. Its chief characters are its broad grains and glumes, the length-breadth index of which provided the main material on which the genetical analyses of Philiptschenko were based. By breeding pure lines of the "Marquis" Wheat (*T. v. lutescens*) and the narrow-grained ordinary Bearded Wheat (*T. v. ferrugineum rossicum*) for several seasons, he first ascertained the range of fluctuating variations in the shape of the grains due to the environment, which served to define at the outset the precise genetical differences between the two pure lines. He then crossed the two pure lines and found the narrow grains and glumes to be incompletely dominant, although it was difficult to distinguish F_1 from the narrow parent P_1. In F_2 pure narrow types were obtained, but in accordance with the principle of the Engledow "shift", the broad "Marquis" type was not recovered until F_3. Owing to frequent overlapping of the various grades it was necessary to test the F_3 plants in F_4, when the situation became clear and the complex results reduced to Mendelian order. In order to secure a complete genic analysis it was necessary to cross the "Marquis" with several other races and varieties of the species, and finally Philiptschenko succeeded in identifying the six pairs of genes which in combination with one another interacted to produce the true "Marquis" wheat. Of these six pairs of genes the basic pair organising broad glumes and grains was the recessive *aa*, the narrow glumes and grains of the common bearded type being organised by the dominant *AA*.

On this recessive base *aa*, five other pairs of modifying genes were found which, interacting with *aa*, produced the true "Marquis" wheat. Of these, two pairs proved to be homozygous dominant genes *BB* and *CC*, while the remaining three pairs proved to be recessive genes *dd*, *ee* and *ff*. The genic formula for the broad grains and glumes of the "Marquis" wheat may therefore be written:

$$[aa + (BB + CC) + (dd + ee + ff)].$$

These exhaustive genetical experiments in wheats by Philiptschenko lead to results which are of considerable technical importance to geneticists and at the same time mark a definite step in the progress of studies in human genetics. First, the experiments have determined the genic formula of the broad-grained super-wheat "Marquis", and also of several different varieties of *compactum* and other wheats used in the tests, as well as of those of the ordinary narrow-grained bearded wheats. As a step to the further improvement of the grain in wheats this knowledge is invaluable. Second, the experiments have interpreted and satisfactorily explained the true nature of the Engledow "shift" met with in many experiments with plants and animals, where the grandparental characters are not precisely recovered in F_2 but are represented by a "shifted" form less in degree or size than in the original grandparent. Third, and perhaps most important of all, Philiptschenko proceeds to compare his results in wheat with the considerable data on the inheritance of musical ability in man that have been accumulated during the last twenty years by himself and others. He finds that if the family data showing high, medium and low grades of musical ability in different individuals be compared with the wheat data showing high, medium and low indices of broadness of the grain, the results of the matings high × high, high × low, low × low, high × medium, low × medium, and medium × medium show ratios remarkably similar in wheat and man. In order to cover the music data Philiptschenko's genic formula for high musical ability may be expressed as follows:

$$[mm + (AA + BB + CC)],$$

where "musical" is represented by the recessive pair of genes *mm* (recessive to the dominant "unmusical" *MM*) plus the three dominant modifying genes *AA*, *BB* and *CC*, each of which produces an equal cumulative effect in the direction of increasing the natural musical capacity represented by *mm*, and which in its weakest expression would be of the constitution *mmaabbcc*. On this scheme the generally accepted hypothesis originally proposed by Hurst (1908) (fig. 156) and supported by data of Davenport (1911), Drinkwater (1916), Philiptschenko (1917), Diakonov and Luce (1922), Stanton (1922), Sacharov (1924), and Mjöen (1926), that in general, basic musicalness is a Mendelian recessive character, is confirmed, while at the same time the contentions of Haecker and Ziehen (1923) and Strogaya (1926), that musical ability is a Mendelian dominant character, are equally confirmed. Indeed it is now clear that general musical capacity and special musical ability are organised by different and independent pairs of genes, the former

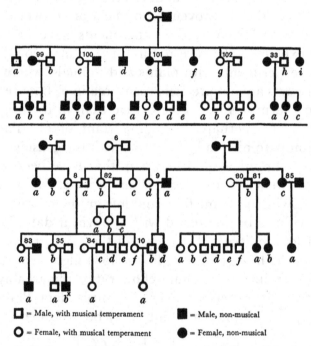

Fig. 156. Pedigrees of musical temperament in Leicestershire families.

being recessive and the latter dominant. As Philiptschenko points out, the valuable data of J. Mjöen (1925, 1926) of the pedigrees of famous musicians, and other musical families in which at least ten children are found, are in close agreement with the genic formula proposed, and there is no doubt that it also covers satisfactorily the other published data.

Philiptschenko does not publish details of the working of his scheme, and it may be interesting to note that on his formula at least seven grades or degrees of musical ability may be expected, which may be set out as follows:

Grade	Class (Phenotype)	Genotype	Dominant genes
7	Classic musicians of the calibre of Bach, Beethoven, Wagner and Mozart	*mmAABBCC*	6
6	Highly talented musicians	*mmAaBBCC*	5
5	Talented musicians of more than average ability	*mmaaBBCC*	4
4	Average musicians	*mmaaBbCC*	3
3	Moderate musicians of less than average ability	*mmaabbCC*	2
2	Indifferent musicians	*mmaabbCc*	1
1	Weakly musical	*mmaabbcc*	0
0	Unmusical	*MmAABBCC* or *MMaabbcc* etc.	

In the musical grades 2 to 6, the five phenotypes may be of twenty-five different genotypes. Thus, grade 6 of highly talented musicians with five dominant genes for ability might be of genotype *AaBBCC*, *AABbCC* or *AABBCc*, while grade 2 of indifferent musicians with one dominant gene might be *Aabbcc*, *aaBbcc* or *aabbCc*, and so on with the other grades except 1 and 7, in which only a single genotype is possible. A striking feature of this scheme, which also is not mentioned by Philiptschenko, is the possibility that the unmusical *MM* and *Mm* individuals may be carrying the dominant genes for high musical ability which are inactive when *M* is present and can only be expressed in the recessive state *mm*. This explains many puzzling cases met with, where heterozygous unmusical parents produce musical children of considerable ability, and it also clears up a number of difficulties found in the Bach and other pedigrees. (Hurst, 1930 A.)

In the absence of suitable experimental psychological investigations we do not know precisely what the cumulative genes *A*, *B* and *C* represent. Being associated with the recessive genes *mm* for musical capacity, they may be simply degrees of musical ability, including capacity for musical expression, musical memory, and sensibility to melody, harmony and rhythm. On the other hand, these independent pairs of genes, which appear to be carried in different pairs of chromosomes in the wheats and rabbits, may represent other psychological characters not necessarily musical, such as high powers of imagination, capacity for emotion and high dramatic sense, or in some cases one might represent simply a capacity for concentration which is necessary for success in any sphere of life. This would increase the number of musical phenotypes and make them qualitative.

In the great musical composers all these qualities seem to be present, and it may be that in them more than four pairs of genes are concerned, just as on another plane the super-wheat "Marquis" has been built up by a peculiar and rare combination of six pairs of dominant and recessive genes. When one considers that the population of Europe is only in part composed of musical people with *mm* genes and that the vast majority of these belong to grades 4 to 1 with from none to three dominant genes for musical ability, the occurrence of a super-musician must be very rare indeed and the extreme rarity of a musical genius is explained. Further, when it is realised that of the total genes carried by an individual only one-half can be transmitted to his offspring, so that a musical genius with, say, six dominant genes for musical ability cannot transmit more than three of them to each of his offspring, it is seen that the chances of a genius producing a genius are very remote though not impossible, as a glance at the genic formulae shows. The same formulae also explain the fact, familiar in all musical pedigrees, that the more gifted the parent the more likely are the offspring to be gifted, and that the higher grades of musicians produce offspring of higher musical grades than the lower grades of musicians do.

The successful application of Philiptschenko's genetical wheat formula to the published data of the incidence of musical ability in human families encourages the hope that by similar methods of research we may be able to approach the most difficult problem of genetics and certainly the one most important to man and the human race, that of the genetical basis of the human mind. Notwithstanding the remarkable output of various school and industrial "intelligence tests" and the numerous studies of genius, insanity and mental defects that have been made, so far little has been done on modern genetical lines to attempt to solve the problems of the genetical basis of ordinary intelligence. It seems likely, however, that a detailed study of the dominant modifying genes associated with the genes for musical capacity may lead eventually to a study of the genetical basis of mind[1]. Both are, in any case, of

[1] Since the above has been set up in type, a genetical formula for the inheritance of general intelligence in man has been constructed and tested on two widely different sets of data

a psychological nature and may in some cases be identical genetically. A study of the genetics of normal intelligence is surely the only sound foundation for future studies of genius, insanity and mental defects. These important problems, however, will not be solved satisfactorily on a simple Mendelian basis, they are far too complicated for that, and it is urgently necessary that the most recent methods and conclusions of genetical research should be applied to these problems. Owing to the complexities of the subject and the many technical difficulties involved, considerable team work is required to bring the research to a successful issue, and close co-operation between experienced geneticists and experimental psychologists is a primary necessity. Once the characters and genes in a series of families are genetically established the statistician can then step in and apply the results to the general population. Once the material has been analysed genetically the use of advanced statistical methods will be of real service and is indeed indispensable.

(Hurst, 1932): (1) 194 Leicestershire families consisting of 388 parents and their 812 offspring individually examined and graded by the author during the last 20 years; (2) 212 Royal families of Europe consisting of 424 parents and their 558 offspring studied and graded by Dr F. Adams Woods (1906). Woods' ten grades of intelligence, which are based on those of Galton (1869 and 1892), are adopted for both sets of data and may be characterised as follows: grade 10 (*Illustrious*), grade 9 (*Eminent*), grade 8 (*Brilliant*), grade 7 (*Talented*), grade 6 (*Able*), grade 5 (*Mediocre*), grade 4 (*Dull*), grade 3 (*Subnormal*), grade 2 (*Moron*), grade 1 (*Imbecile*).

The genetical formula in its most heterozygous form is:

$$Nn + (Aa + Bb + Cc + Dd + Ee)$$

where N is a major gene determining mediocre intelligence of grade 5, and $A...E$ and $a...e$ are minor increaser and decreaser genes respectively, acting only in the presence of nn. In the presence of the major gene N the minor increaser and decreaser genes are inactive and inhibited.

729 genotypes are possible, e.g. grade 10 (*Illustrious*) is *nnAABBCCDDEE*, grade 0 (*Idiot*) is *nnaabbccddee* and so on for grades 1–9, each dominant minor gene increasing the grade by 1 and each recessive decreasing it. Grade 5 (*Mediocre*) may be, e.g. *nnAaBbCcDdEe*, etc., or (*NN* or *Nn*) + any minor genes. Since the minor genes, so far as the evidence goes, are quantitative and not qualitative, only seven gametic types are possible, and on this basis an Expectation Table has been made which predicts the grades of intelligence expected in the offspring of the matings of all the grades with one another.

The formula applied to the Leicestershire and the Royal families shows that out of 1370 offspring 1343 (98·1%) are of the grades expected, while 27 (1·9%) are exceptions. A large proportion of the exceptions occur in the offspring of pathological parents of the highest grades 10 and 9, pointing to a pathological disturbance due to the environment or to other genes.

The next step is to use this genetical formula as a working hypothesis and to test it further and more fully in other families and populations.

CHAPTER XIV

SEX

WE have seen how the presence of different chromosomes in the different sexes led to the discovery of the genes being located within the chromosomes. Since those early days much work has been done on the inheritance of sex-linked characters and on the heredity of sex itself, and more has been learned of the workings of the genes by this means, not only their positions in the chromosomes but also their actual physiological effects, than by any other. Here we have two definite clear-cut categories—male and female. In all the higher forms of life they are definitely distinguishable from one another by structure and by function, and in many cases they are also distinguished by other striking characteristics, such as different arrangements of hair and feathers, different colourings and a different psychology. The study of these differences from various standpoints has provided valuable information concerning genetics and evolution.

It was early discovered that certain characters were inherited in a curious manner, all the sons bearing one contrasted character and all the daughters another. In cultures of the Fruit Fly (*Drosophila*), it was found that if one mated a female with white eyes with a male with red eyes, in the resulting progeny all the sons were white-eyed while all the daughters were red-eyed, the characters having crossed over to the opposite sex. On mating these together, however, white-eyed males and females and red-eyed males and females were obtained in equal numbers. If, however, we reverse the procedure and cross a red-eyed female with a white-eyed male, all the progeny male and female have red eyes, and when these are bred we get the Mendelian ratio of 3 red-eyed individuals to 1 white-eyed, but in this case all those with white eyes are males and *there are no white-eyed females*; the white-eyed grandfather has passed on the gene for white eyes to half his grandsons but only in a concealed state to his granddaughters. Such behaviour as this was very puzzling,

but once the genes were connected with the chromosomes these complexities became clear and simple.

As we have seen, the male and female *Drosophila* have different
chromosome complexes (p. 38). The female has two pairs of long
curved chromosomes, one pair of tiny chromosomes and two straight
X chromosomes, while the male has the same except that in place of the
two X chromosomes he has only one X chromosome and another longer
J-shaped chromosome known as the Y chromosome. These X and Y
chromosomes have the power, together with the balance of the other
chromosomes, to determine the different sexes, the females having two
X chromosomes and the males an X and a Y.

Many experiments have shown that in *Drosophila* the Y chromosome,
though paired with the X, carries no genes corresponding to those in the
X chromosome, so that as far as the action of the genes in the X chromosome is concerned the Y might as well be non-existent. Re-examining
the curious case of criss-cross inheritance of eye colour in *Drosophila* on
this basis, we find that there is a simple explanation. The gene for red
eyes is carried by the X chromosome but there is no corresponding gene
for it in the Y chromosome, so that the male has only one chromosome
carrying it, while the female, with two X chromosomes, has two. White
eye colour is recessive to the red, from which it arose as a mutant. When
we cross the white-eyed female (with two X chromosomes carrying
white) with a red-eyed male (with one X carrying red and a Y without),
in the progeny we get one white X from the female meeting the red X
from the male, giving 50 per cent. females with red eyes, since red is
dominant and two X's make a female, while in the other 50 per cent.
we get the white X from the mother meeting the empty Y from the
father, giving white-eyed males, since there is nothing in the Y to
obscure the recessive white and the XY combination gives a male. Thus
we see how it is that we get one-half *red-eyed females* and one-half *white-
eyed males* from a cross between a *white-eyed female* and a *red-eyed male*,
a result which at first sight appears to be paradoxical. A study of
fig. 157 shows how segregation of the chromosomes will give the proportions in the next generation and also in the reverse cross, in which
an entirely different result occurs. The black bars stand for the X chromo-

somes bearing the gene for red eyes, the white ones for those bearing white. We now see at a glance how the white-eyed male will produce only white-eyed grandsons and no white-eyed granddaughters. Since he only bears one X chromosome and white is a recessive character it is always masked by the red X chromosome from the female in the grand-daughters, and it is only when it meets the empty Y chromosome that it can appear, and this combination will of course produce only males.

In certain kinds of fish—the Millions Fish (*Lebistes reticulatus*) and a small ornamental fish of Japan (*Aplocheilus latipes*)—Schmidt, Winge

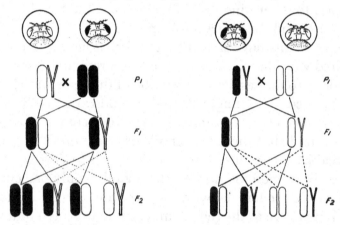

Fig. 157. Diagrams representing the sex-linked inheritance of red and white eye colour in the Fruit Fly (*Drosophila*). On the left a white-eyed male × red-eyed female; on the right a red-eyed male × white-eyed female. The individuals with two bars (which stand for X chromosomes) are female, those with a Y are male; black bars carry the "red" gene, while white bars carry the "white" gene. (After Morgan.)

and Aida have found that the Y chromosome is not without varietal genes as it is in *Drosophila*. In *Lebistes* the females are all greyish green in colour, while the males have brightly coloured spots, these being differ-ent in different races. No matter how these different races are bred and inter-bred the males always transmit their distinguishing spots to the sons only, while all the female progeny are the same plain colour as the mother. This can be explained if the genes for these male character-istics are carried by the Y chromosome, which is present only in the males.

In all organisms that have males and females living as separate individuals we may expect to get these curious types of inheritance closely bound up with sex. In man we have the phenomena of Colour-blindness and Haemophilia (Bleeders), both of which affect more males than females. These are recessive characters which behave in heredity like the white eye colour in the male in *Drosophila*; they are inherited through females, but rarely appear in that sex. Observations show that a colour-blind father and a normal mother have all their sons and daughters normal (fig. 159). All the daughters can transmit the defect but none of the sons can do so. If the daughters marry normal men the defect will be transmitted to one-half of their children of both sexes, but only the sons will be colour-blind. On the other hand, if the daughters

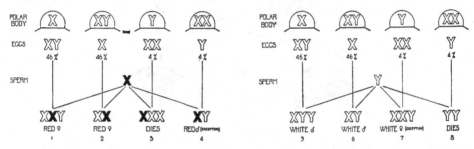

Fig. 158. The fertilisation of a female *Drosophila*, having two *X* chromosomes (both carrying the recessive white eye colour) and a *Y*, by a normal red-eyed male. Left, fertilisation by the sperm, carrying the *X*, of the four types of eggs produced, and right, by the *Y*. The offspring with three *X*'s or with 2 *Y*'s die and the usual sex-inheritance is altered since the female carries a *Y* (normal females having only two *X* chromosomes) and this, meeting the male *X*, gives a red male while the *Y* from the male meeting the two white *X*'s from the female gives a white female. (After Morgan.)

marry colour-blind men, one-half of their daughters and one-half of their sons will be colour-blind. In such cases the affected sons and daughters will carry on the defect, and if an affected daughter marries a colour-blind man all their children will be colour-blind. The un-affected daughters will still carry on the defect but the unaffected sons will not. The explanation of all these complicated facts is to be found in the simple mechanism of the *X* and *Y* chromosomes (fig. 160). The gene for colour-blindness is located in the *X* chromosome and not in the *Y* chromosome. If **X** represents the sex chromosome carrying the gene for normal colour perception and *X* the sex chromosome carrying the gene for colour-blindness, then all males are either **X**Y normal or XY colour-blind. Females, on the other hand, are of three kinds: **XX** normal, X**X** normal (carrying the defect) or XX colour-blind. One dose of the gene makes a man colour-blind, but it takes two doses of the gene to make a woman colour-blind. The Mendelian segregation of the X, **X** and Y chromosomes in the reduction divisions of the eggs and sperms and their random matings in fertilisation provide a simple explanation of the complicated facts of the inheritance of colour-blindness in man (fig. 159).

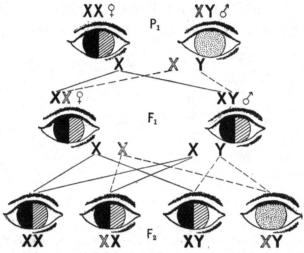

Fig. 159. Diagram representing the genetics of colour-blindness in Man. In this case the original mother was normal and the father colour-blind. The eyes, which are half black and half barred, are normal, the stippled ones colour-blind; the black X stands for the chromosome carrying the normal gene, the stippled X for the one carrying the recessive colour-blind gene. In F_1 we get normal males and normal heterozygous females, in F_2 we get normal males and females, normal heterozygous females and colour-blind males in equal numbers. (After Morgan.)

In poultry this sex-linked type of inheritance (fig. 167) has become of practical importance, since by crossing certain breeds one may get the resulting day-old chicks of two colours, the males one colour, the females another, and it is thus possible to pick out the pullets as soon as hatched, which is sometimes a considerable asset to the breeder. Prof. Punnett has recently raised a new breed which presents this phenomenon within itself without recourse to cross-breeding. In this breed, called` the Silver Cambar, the pullets can be distinguished from the cockerels at hatching time.

Not only have the inheritance of characters borne by the sex chromo-

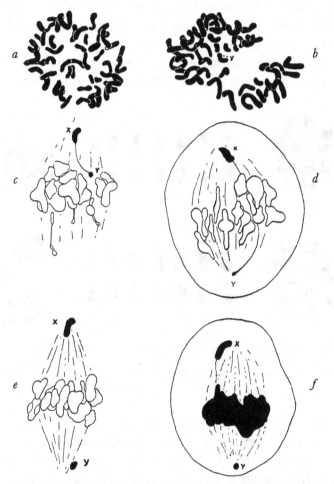

Fig. 160. Chromosomes in Man. *a, b,* 48 somatic chromosomes; *c, d, e, f,* reduction divisions in the male, showing the *X* and *Y* sex chromosomes reducing first before the autosomes. Thus in man the male is digametic for sex. (After Painter.)

somes been studied, but a good deal of work has been done on the differentiation of sex physiologically. We have seen how work on triploid *Drosophila melanogaster* has proved that this depends, in this species at all events, not so much upon genes borne by these chromosomes as on the balance of the chromosomes as a whole, different balances causing males, females, inter-sexes, super-males and super-females. Although no varietal genes have been discovered in the Y chromosome in this species, yet apparently normal males which have occurred from time to time bearing the X chromosome alone with no Y have been completely sterile, thus showing that although the Y chromosome has no apparent effect upon the varietal characters of the individual yet it possesses something which affects fertility, since in its absence males are infertile.

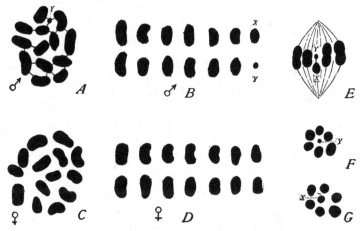

Fig. 161. XY sex chromosomes in the hemipter *Lygaeus turcicus* in which the male is digametic for sex. Above, the male ($\male$) chromosome complex with the X and Y chromosomes; below, the female ($\female$) with the two X chromosomes. F and G show the two kinds of gametes containing X and Y respectively. (After Wilson.)

A great deal of work has been done by Goldschmidt on the Gipsy Moth (*Lymantria*) bearing upon this balance of the sexes. There are numerous varieties of this moth, European, Asiatic and American, and by inter-breeding the different races it has been found possible to produce any type of inter-sex at will (fig. 162). If these moths breed naturally, each variety by itself, everything proceeds normally, males and females occurring in the usual way. If a Japanese male is mated with a European female normal male offspring occur, but all the female offspring have certain male characteristics and may be regarded as inter-sexes. In some cases it was possible to breed from these inter-sexes, and on crossing them with their normal brothers the next generation showed ordinary Mendelian segregation, one half of the females produced being normal and the other half inter-sexes. If, however, the cross is reversed and a European male is mated with a Japanese female, normal males and females arise in the first generation, but on breeding

Fig. 162. Inter-sexes in the Gipsy Moths (*Lymantria*). 1, 2, the male and female of *L. dispar*; 3, 4, male and female of *L. japonica*; 5–16, hybrids between the two showing various types of inter-sexes. (After Goldschmidt.)

from these a certain proportion of *males* arise with female characteristics, that is to say, in this case it is the males which are inter-sexes. Further work showed that among the many races of this moth all degrees of sex "strength" or "weakness" relative to each other occur, each race breeding normally itself but producing different results on being out-crossed to other races. Thus if a race graded as producing "very strong" males is mated to the females of a race with "weak" femaleness, all the offspring are male in their type of organisation. If a "strong male" is mated to a "weak female", half the offspring are normal males, the other half female inter-sexes, and so on. In this way by suitable inter-racial breeding it is possible to get every type of inter-sex, from one almost male to one almost female, in addition to the normal males and females. Giving relative values to the different strengths of maleness or femaleness within the various races, Goldschmidt postulates that

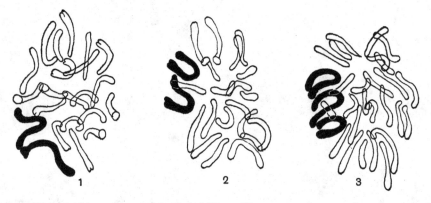

Fig. 163. Polyploid inter-sexes in the Sorrel Plant (*Rumex acetosa*). 1, 2, chromosomes of two different triploid plants with 22 chromosomes (= 18 autosomes + 2 X + 2 Y) and, 3, of a tetraploid plant with 29 (= 24 A + 3 X + 2 Y). The normal diploid plants of either sex have (male) 15 chromosomes (= 12 A + X + 2 Y) or female 14 chromosomes (= 12 A + 2 X). Triploid and tetraploid inter-sexes are found occasionally and have evidently arisen by unreduced male or female gametes. It will be observed that in this species the autosomes are euploid, 12, 18 and 24, while the total chromosome numbers are aneuploid, 14–15, 22 and 29. (After Ono and Shimotomai.)

if the unit representing maleness is greater than the unit representing femaleness the result of the cross is a male in its main characteristics, but if the value given to the female be the greater the result will be female.

Since *Lymantria* has a relatively high chromosome number (62 somatic chromosomes), it seems not unlikely that this curious quantitative effect of the genes determining the sexual characters may be due to a condition of masked polyploidy, which would involve the presence of more than one set of sex chromosomes, the different strengths of the sexes in the different races being the manifestation of differing combinations of Z and O chromosomes, some races having more male elements while others have more female. We have seen how intermediate types arise in polyploids due to the interaction of the different combinations of genes when more than the one pair is present. Here we find that the genes may be present in different "strengths" giving varying results. In *Lymantria* the variation is apparently due to different rates of development of the male or female differentiation, so that in

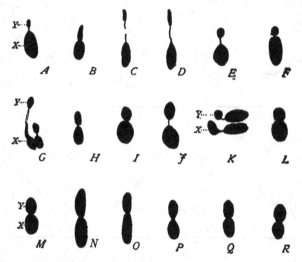

Fig. 164. Examples of the *XY* chromosomes in various insects. *A*, the beetle *Trirhabda*; *B, C, D*, the hemipter *Nezara viridula*; *E*, the hemipter *Lygaeus turcicus*; *F*, the beetle *Chrysocus*; *G*, the hemipter *Notonecta indica*; *H*, the hemipter *Thyanta custator*; *I*, the hemipter *Euschistus fisilis*; *J*, *Lygaeus bicrucis*; *K*, the mosquito *Anopheles* with the *X* and *Y* joined to a pair of autosomes; *L* to *R*, various Hemiptera: *M*, *Mineus*; *N, O*, *Nezara hilaris*; *P, Q, R*, *Oncopeltus fasciatus*. (After Wilson.)

cases of late development the opposite sex becomes ascendant, overcoming the one to be normally expected from the chromosome complex.

From much work done on various animals, birds and insects, it is evident that although the actual sex of the individual depends on the balance of the sex chromosomes with the autosomes, yet occasionally external influences may cause disturbance of this, in some cases even a complete reversal from one sex to another. When fertilisation takes place the resultant individual is, so far as its chromosome complex is concerned, either male or female and, with the exception of the rare

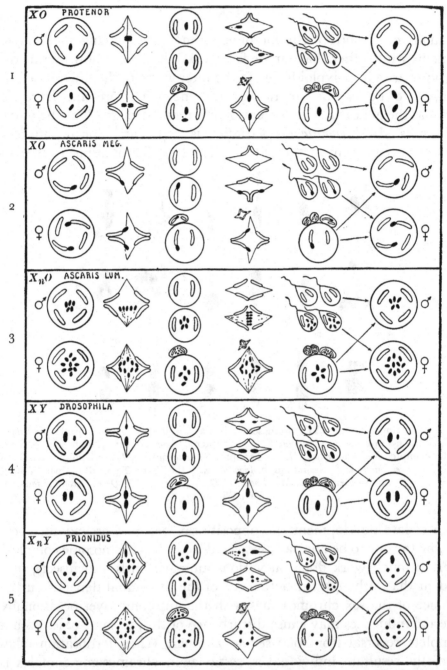

Fig. 165. Diagram showing the different types of sex chromosomes and their behaviour in animals in which the male is digametic for sex. 1, *XO* type; 2, *XO* type in which *X* is joined to an autosome; 3, X_nO type in which *X* consists of several elements; 4, *XY* type; 5, X_nY type in which *X* is complex. (After Sharp.)

cases of chromosomes being eliminated, will remain so and breed in this way whatever may happen to its other characteristics. In many animals the primary germ-cells of the individual are laid down from the beginning, and any change in the body which bears them will have no effect on them except under rare circumstances, so that an individual which starts as a male with XY chromosomes will remain so, as far as its germ-cells are concerned, giving off X and Y gametes in equal numbers, while a female XX will produce all gametes carrying X. When the body becomes differentiated certain organs arise which, functioning, produce characteristic secretions which have great influence on the whole growth of the body. Thus from the male organs a certain kind of "sex-hormone" is secreted which controls the growth of male characteristics, while in the females a different kind of hormone arises which produces female characteristics (fig. 166). Normally all proceeds well, the fertilised egg, with its male or female complex of chromosomes containing the genes to produce the differentiation peculiar to the species, divides and the cells keep multiplying, assuming gradually the form of the adult body, laying down the different organs and tissues with their varied functions which are to carry on the specialised work allotted to them. Sometimes, however, accidents occur and the balance of development is upset; man may for one reason or another produce this artificially or a disease may supervene which has the same effect. Much interesting work on this has been done by Crew and others in poultry. From time to time hens have been known which as they get older take on male characteristics, assuming the feathering, spurs, wattles, voice and behaviour of a cock bird. One such hen, a Buff Orpington of $3\frac{1}{2}$ years, the mother of chickens and a good layer, assumed all the attributes of a cock, crowing lustily and even in the end becoming the father of chickens. Post-mortem examination proved that the ovary had been almost entirely destroyed by tuberculosis. Its disappearance and the consequent cessation of its functioning had so upset the general metabolism of the bird that it had caused the growth of male tissue, so that eventually the bird was able to function as a male and became male in appearance. This complete sex reversal due to diseased ovarian tissue is indeed remarkable, and shows the strength of the secretions which arise on the

Fig. 166. The influence of sex hormones in Poultry. On the left all the birds are male in their gametic constitution, those on the right are all female. 1, normal cock; 2, a capon (cock with testes removed) with male plumage but neuter comb and instincts; 3, male with testes removed and ovaries engrafted, with female comb, plumage and instincts but size and carriage of male retained; 4, normal hen; 5, a poulard (hen with ovaries removed), resembles the capon except in size; 6, hen with ovaries removed and testes engrafted, which has become male in all respects except build and size. (After Finlay.)

differentiation of even one part of the body and their far-reaching effects, indicating that it will not take a very large alteration of the gene or chromosome complex to produce a profound difference in the resulting individuals. A good deal of work has been done removing testes and

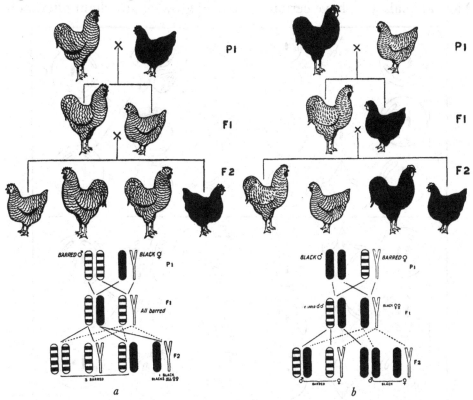

Fig. 167. Sex-linkage in Poultry in which the female is digametic for sex, thus reversing the conditions found in *Drosophila* (fig. 157). *a*, barred cock × a black hen gives all the F_1 barred whatever the sex while in F_2 only half the granddaughters exhibit the recessive sex-linked character, the grandsons all being homozygous or heterozygous barred; *b*, the reverse cross, which gives an F_1 with barred sons and black daughters and an F_2 with equal numbers of barred and black sons and daughters. In the diagram of the sex chromosomes below the black and barred represent the Z chromosomes and the white forks the W chromosomes or its absence O. (See figs. 170 and 171.) (After Crew.)

ovaries from cocks and hens and engrafting male tissue in place of female and *vice versa*, all showing the same results, that once the various organs become differentiated the secretions given off by them take charge, as it were, and any alteration by accident or design will cause a more or less complete change of function which will entirely alter the appearance of the individual. In man we are familiar with the unfortunate effect of a deficiency or excess of secretion of the thyroid gland, either of which will produce a disastrous effect upon the whole body. Most animals, too, have certain centres of growth, any slight alteration

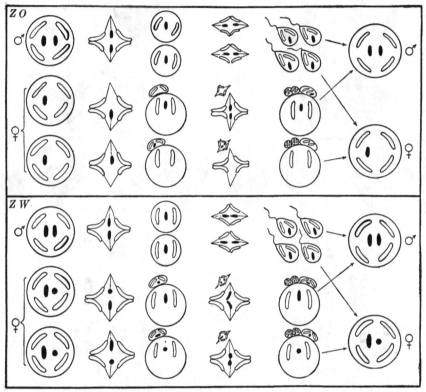

Fig. 168. Diagram of the ZO and ZW type of sex chromosome behaviour in animals in which the female is digametic for sex. (After Sharp.)

to which will cause great differences of bodily size and shape, so that a mutation of one of the genes controlling one of these centres may have a profound influence upon the whole race.

An interesting experiment by Danforth shows the different modes of development of different parts of the body. In some breeds of domestic fowls the feathers of the cock are loose and unbarbed while those of the hen are tight and stiffly barbed. Danforth took a small piece of skin from the back of a male chick of this type with barred feathers and grafted it on the back of a female chick with plain coloured feathers. As the chick grew into a hen the piece of grafted skin grew with it and retained the original barred colour pattern of the male chick from which it was taken, but the feathers did not retain the male characteristics. They were stiff and barbed like the other normal feathers of the hen (fig. 169), thus showing that, while the colour characteristics are locally differentiated, the feather formation is controlled by the internal secretions of the individual as shown by the alteration of the feathers in the hens which become cocks. Further experiments on these lines should give us more insight into the physiological action of the genes. Considerable work has been done on sex reversal in other animals and the differentiation of sex, but since much of it has not been associated with cytological observations, interpretation is often difficult, and no useful end would be served by referring to the results obtained.

In addition to their genetical behaviour there are also many points of interest in the sex chromosomes themselves. We have already seen that (so far as examined) in mammals, fishes and many insects it is the male which is digametic and determines the sex of the offspring, while the female is homogametic for sex, her sex chromosomes being similar. In some cases the male has only one sex chromosome—the X—while in others it has two unlike chromosomes—the X and Y. In *Drosophila* we probably have what may be considered an intermediate type, for although the Y is present it apparently contains no genes which influence the varietal characters of the fly, since males without a Y are externally the same as those which possess it. One all important function it possesses, however. Since flies without it are sterile, presumably it contains genes which are essential to fertility, so that the Y is necessary to the continuance of the race. If the few necessary characters contained in it should by translocation become attached to another chromosome, then the Y might drop out without injury to the individual. Individuals without the Y chromosome are said to have the XO type of sex inheritance, and the males have a chromosome less in their body cells.

In Birds and some Moths (*Lepidoptera*) it is the female which is

A	*B*	*C*	*D*

2

Fig. 169. 1, a plain-coloured hen upon which was grafted, when a chick, a piece of skin from a male chick with barred plumage. The male colouring and pattern have been retained but the type of feather is female. 2, left, feathers from skin grafted on male hosts: *A*, from a black female, *B*, from a barred male; female feather (*A*) has retained its black colour—a sex-linked character, but is typically male in shape—a secondary sexual character; right, two feathers from the fowl depicted in 1: *C*, a normal feather, *D*, a feather from the grafted male skin which has taken on the female characteristic shape but retained the barring of the male *B*. (After Danforth, *Journal of Heredity*.)

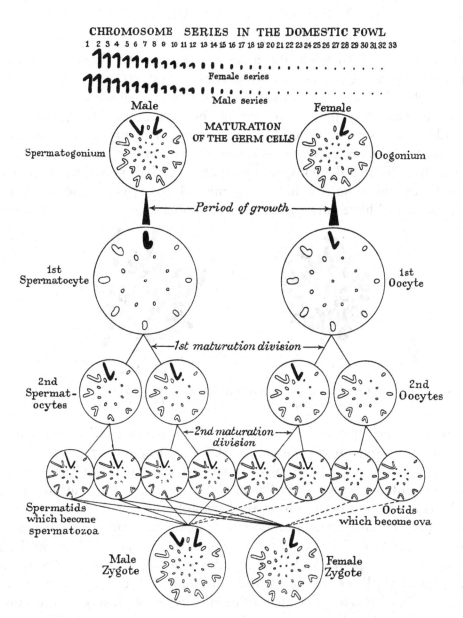

Fig. 170. Diagram showing the formation of the germ-cells in the cock and hen respectively of the domestic Fowl (*Gallus*). The black chromosomes are the sex chromosomes, the male showing two Z chromosomes and the female only one, although some workers have identified a small W chromosome also (see fig. 171). In either case the female is digametic for sex. (After Hance.)

digametic for sex, having two unlike chromosomes (ZW) or only one sex chromosome (ZO), while the male is homogametic, having two like chromosomes (ZZ); this is known as the ZW and ZO type of inheritance (fig. 168). Further variations of these types arise. In some cases the sex chromosome may join on to one of the autosomes during divisions, while in others it is made up of more than one element (figs. 172 a and b). In the male of *Ascaris lumbricoides* the X chromosome consists of no less than five distinct elements, all of which go to one pole during the reduction division, so that half the gametes have 24 chromosomes while the other half have only 19. In some cases these compound sex chromo-

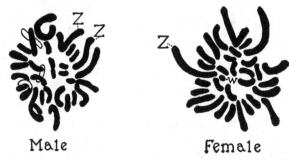

Male   Female

Fig. 171. Chromosomes in male and female Fowls showing the ZZ and ZW chromosomes respectively. (See fig. 170.) (After Shiwago.)

somes remain separated throughout, at others they come together and form large irregular chromosomes. (Cf. figs. 165 and 172.)

In some dioecious plants with separated sexes identical mechanisms for sex differentiation have been discovered (fig. 173), although in plants a vast number of forms are hermaphrodite or monoecious, that is to say, the same plant produces both male and female gametes. This may be effected in more than one way—the flowers may have anthers containing the pollen and male gametes together with ovaries containing the egg-cells or female gametes (hermaphrodite), or these may, as in the Hazel (*Corylus*), occur on different types of flowers on the same plant, the catkins being the male flowers, the tiny red tassels the female (monoecious). In other cases the male and female flowers are on entirely different plants (dioecious) in the same way that different sexes appear in different individuals in the majority of animals, though here

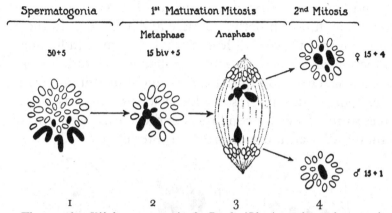

Fig. 172 *a*. The complex *XY* chromosomes in the Beetle (*Blaps*). 1, shows the somatic chromosomes, 30 autosomes + 5 sex chromosomes; 2, the preparation for reduction for the gametes, 15 pairs + the 5 sex chromosomes joined together; 3, the reduction division, the *Y* chromosome goes to one daughter cell, the 4 *X* chromosomes to the other; 4, the gametes each with 15 autosomes, but one with 4 sex chromosomes the other with only 1. (After Wilson.)

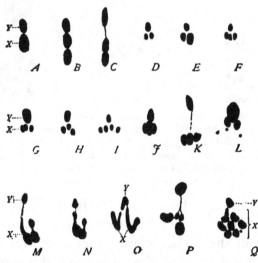

Fig. 172 *b*. Different types of the compound sex chromosome *X* and *Y*. *A, Thyanta custator*; *B, C, Thyanta calceata*; *D, Fitchia*; *E, Conorhinus*; *F, Sinea*; *G, Pselliodes*; *H, Prionidus*; *I, Sinea rilei*; *J, K, Gelastocoris*; *L, Acholla multispinosa*; *M, N, Notonecta indica* (all these are Hemiptera); *O,* Mantid Orthopteran *Tenodera superstitiosa*; *P,* Beetle *Blaps lusitanica*; *Q,* Nematode Worm *Ascaris incurva*. (After Wilson.)

too we find hermaphrodite forms, such as the snail, in which the same individual may function both as a male and a female. In the oyster we get another form of sexual differentiation, in which individuals apparently function alternately as males or females at different periods of their existence. The whole question of sexual differentiation is extremely intricate, being bound up in some lower forms with seasonal variations of temperature and other conditions and even higher forms are liable to fluctuations by accident or artificial interference.

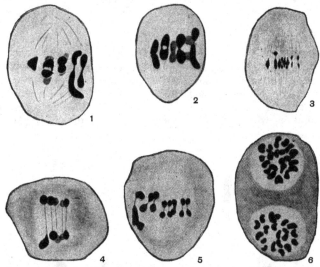

Fig. 173. Sex chromosomes in flowering plants. 1, gametic division in *Melandrium album* (White Campion) showing unequal *XY* pair; 2, the same in *Humulus japonicus* (Japanese Hop) and, 3, *Populus balsamifera* (Balsam Poplar). 4, later stage in *Valeriana dioica* (Marsh Valerian) showing the *X* and *Y* separated; 5, *Rumex acetosella* (Sheep's Sorrel) showing a complex *M* separating from the *m* chromosome; 6, later stage of same showing 21 chromosomes in one cell and 20 in the other. (After Sharp.)

As we proceed through the various Phyla of animals and plants we find all degrees of reproduction between sexuality and asexuality. In some cases, even in the higher plants, there are many species which reproduce without any sexual fusion at all, and this condition becomes more common as we move lower down the scale of life. Between the two extreme types we find almost every variation. In other cases, as in the Fungi, there is no definite external differentiation of sex, but two or more strains occur, outwardly the same, which act as if definitely male and female. If we term these A and B strains, respectively, we find that though individuals from the A strain will fuse with individuals from the B strain, yet A will not fuse with A, nor B with B. Here we probably meet with the most primitive occurrence of sexual differentiation, too primitive to give any external characteristics, but yet separated from one another by an internal constitution which necessitates fusion with members of a different strain. So we find, running parallel with the evolution of the different forms of animals and plants, the evolution of sex. By the tremendous impetus which the presence of its mechanism gives to the introduction and perpetuation of new characters, it has itself been a most potent factor in evolution.

CHAPTER XV

VIRGIN BIRTH

VIRGIN birth is a frequent phenomenon in certain plants and animals, from the dandelion of the fields to the green fly in our gardens. That is to say, individuals are produced from seeds and eggs without fertilisation, they have a mother but no father. In animals, virgin birth is known scientifically as parthenogenesis, while in plants it is termed apogamy or apomixis. Many of the lower organisms reproduce in this way, but of late years the same mechanism has been discovered in many of the higher plants. We are all familiar with the sight of a field of dandelions in bloom. There is something very striking about their *alikeness*, each plant is so very similar to every other plant, the leaves form just the same sort of rosette of just the same colour and the flowers are identically the same. The explanation of this extraordinary lack of variation is that they are reproduced apomictically, that is, their seeds are not fertilised but develop automatically without the introduction of any male pollen. Since no male nuclei with their contained chromosomes and genes come in from another parent these plants are identical in all respects with the mother plant, in fact they can hardly be regarded as new individuals, but more as vegetative clones such as we get when we take cuttings, buds or grafts. Technically however, they are true seedlings arising from seeds developed from the ovules of the mother parent.

The great majority of our English wild roses—those known as dogroses, or briars—also reproduce themselves apomictically like the dandelions and hawkweeds. As we have seen, they have a remarkable provision for reproducing themselves in the sexual manner, but they are also facultatively apomictical, that is to say, if they do not get fertilised by any good pollen they will carry on without it and reproduce apomictical seeds without fertilisation from the alternative embryo-sacs which contain nuclei with the full somatic number of chromosomes (Hurst, 1931).

Quite often, but not always, apomixis and parthenogenesis are associated with polyploidy. The most striking cases of this occur in animals, where, as we have seen, the perpetuation of polyploid varieties depends principally on this factor. In the Brine Shrimp (*Artemia salina*) (fig. 174) the diploid species reproduces in the usual sexual manner while the tetraploid variety reproduces parthenogenetically. The Isopod *Trichoniscus provisorius* has a triploid variety with 24 chromosomes which is distributed over Northern and Central Europe and is exclusively parthenogenetic, while the diploid species is bisexual and Mediterranean. The Ostracod *Cypris* has a triploid species *fuscata* with 24 chromosomes and a similar distribution and reproduction, while the

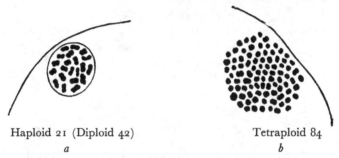

Haploid 21 (Diploid 42) Tetraploid 84
a b

Fig. 174. The chromosomes of the two races of the Brine Shrimp (*Artemia salina*). *a*, the diploid race with 42 chromosomes, a gamete showing 21 as the reduced number; *b*, the tetraploid race, a somatic cell with 84 chromosomes. (After Artom.)

diploid species *Notodromes monacha* has 16 chromosomes and is bisexual and distributed in North Africa.

Two species of the Psychid (*Solenobia*) have tetraploid varieties which are parthenogenetic while the diploids are bisexual. The Cladoceran (*Daphnia pulex*) has a hexaploid variety with 24 chromosomes usually confined to the arctic regions; this polyploid is parthenogenetic, with the males unknown; the diploid has 8 chromosomes and is both sexual and parthenogenetic. The same thing applies to the plant *Alchemilla* (Lady's Mantle), in which the tetraploid form reproduces apomictically. Some other plants also show an association of apomixis and polyploidy and it seems probable, in some cases at all events, that hybridity is a third factor. The association between the two former does not always hold, however, since many polyploid species only reproduce sexually (e.g. *Rosa*), while some diploid species reproduce apomictically, especially in insects.

Quite a number of small animals reproduce in this way, such as *Aphis* (the green fly on roses and other plants), which go on from generation to generation without sexual reproduction. Generations, however, do occur in the autumn in which they become sexual (figs. 175, 176).

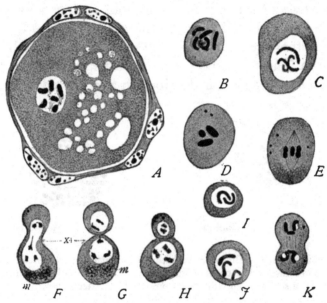

Fig. 175. The two different modes of reproduction in *Aphis saliceti* (Willow Green Fly). *A*, parthenogenetic egg with full diploid number of chromosomes (6); *B–K*, the reduction division in the male gametes showing (*I* and *J*) the two types produced, half of them carrying two chromosomes and the other half three. The means by which one chromosome drops out of the parthenogenetic females to give rise to the male with five chromosomes is somewhat obscure, though it is probably dropped out in a division. In the male gametic divisions the gametes minus a chromosome are much smaller than the others and degenerate so that only female-producing ones function, thus returning to the parthenogenetic cycle again. (After Baehr.)

In all the cases mentioned and in the greater part of the known cases the apomictical eggs are formed without any reduction in the number of chromosomes, so that the female germ-cells contain the same number of chromosomes as the body-cells, with no addition of any chromosomes from the male, and produce an individual perfect in all its parts with a full complement of chromosomes. Since usually there is no reduction of the chromosomes there is likewise no segregation, and of necessity every kind of gene that is carried by the mother must also be received by the offspring, so that, except in those rare instances when mutations arise, they are identical in all respects and may be regarded merely as seed clones from the parent stock. Plants and animals of this type can

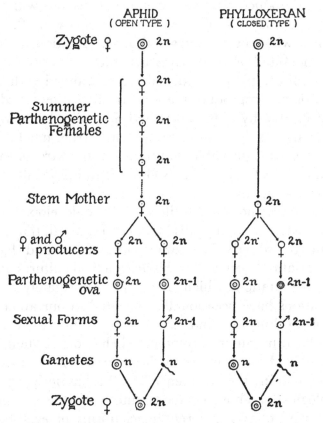

Fig. 176. Diagram illustrating two types of alternate parthenogenetic and sexual generations, the Aphids, where there is a series of parthenogenetic female-producing generations followed by a sexual one, and the Phylloxeran, where there are only two parthenogenetic generations. (After Wilson.)

go on indefinitely without any change, and many of them may be regarded as blind alleys in the evolutionary scheme or as permanent specialisations which occupy a special niche.

They have one great advantage, however, over ordinary sexual plants and animals. In cases where they do mutate the change is at once made permanent and does not run the risk of being obliterated by crossing with other forms carrying the normal character. Ostenfeld (1921) and Rosenberg (1906) have done much interesting work on the Hawkweeds (*Hieracium*). Some of these, like *Rosa*, are facultatively sexual and can produce two types of embryo-sacs, one with the reduced chromosome numbers for fertilisation and one with the unreduced number which will produce seeds without fertilisation. Attempts were made by Mendel and others to hybridise these with other species, and though many of the plants produced were identical with the mother and had evidently reproduced apomictically, a few seedlings arose which were clearly hybrids. These hybrids, also being apomictical, became permanently fixed at once and bred true, much to Mendel's disappointment. It is probable that much of the extreme variability in *Hieracium* and in the *Caninae* roses is also due to this faculty for hybridisation followed by apomixis.

In both *Hieracium* and *Rosa*, fertile seeds have developed from flowers from which all stamens and pistils had been removed, thus taking away any possibility of ordinary fertilisation. Ostenfeld found that occasionally chance variations occurred in his apomictical cultures and he attributes these to mutations which bred true immediately by apomixis. Similar variations have occasionally appeared in our apomictical cultures of *Rosa* at Burbage and Cambridge. Since these apomictical *Hieracia* also have irregular chromosome behaviour, Ostenfeld thinks it probable that a chromosome may have gone astray in the somatic divisions, thus causing the divergence, but this has not yet been demonstrated cytologically. Thus in these facultatively bisexual and apomictical species we get two different mechanisms of evolution, (1) by hybridisation, the results of which become fixed by apomictical reproduction, and (2) by mutations, which also become permanently fixed immediately. Given that these species are capable of occasional hybridisations or have an instability of the genes and chromosomes, new species and varieties will arise and achieve immediate permanence without any of the difficulties arising in sexual forms owing to sterility or the presence of predominating genes. We have seen that the numerous species and varieties of the *Caninae* roses must have originated in this

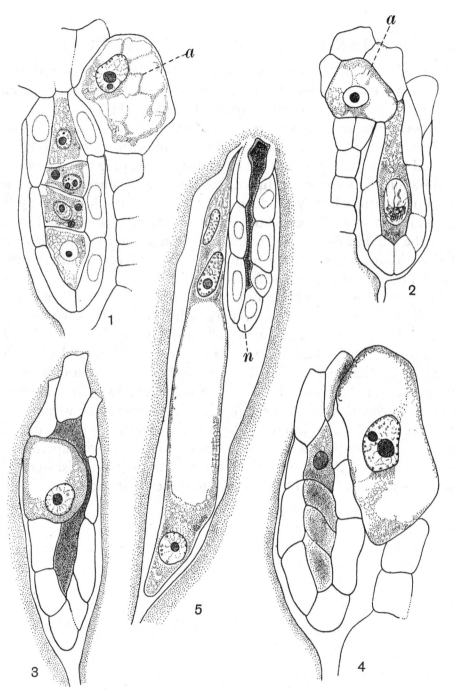

Fig. 177. The development of sexual and apomictical embryo-sacs side by side in *Hieracium* (Hawkweed). 1, 2, 3 and 4 show a cell (*a*) of the ovarian tissue enlarging to form an apomictical sac with the full somatic number of chromosomes and the normal sexual cells with the reduced number in the centre. These have degenerated in 3 and 4. In 5 the apomictical cell has grown into a young embryo-sac while the sexual one to the right (*n*) is degenerate. (After Rosenberg.)

way as far back as the Mindel glaciation of the Pleistocene Period, since the accumulated geological, palaeontological, cytological, genetical, taxonomic and geographical evidence all confirm this theory.

In bees, ants, and wasps a curious mixture of sexual reproduction and parthenogenesis is characteristic. The eggs undergo a reduction as in normal bisexual animals, but are capable of developing with or without fertilisation. If they are fertilised, females are produced, but if they remain unfertilised, they produce males (drones) which have only the haploid number of chromosomes in their body-cells, while the females have the full diploid complement. Here we have much scope for variation in the reduction of the chromosomes in the female and the introduction of the male sperm at fertilisation.

Large numbers of the lower organisms reproduce asexually in part or entirely, since many one-celled and pre-cellular organisms carry through their divisions and form two new cells without any form of reduction and fusion. Others, higher up, behave like the aphids and go on for many generations without fertilisation, and then one generation occurs in which they reduce their chromosome number and two cells fuse, giving a new individual, or the apomictical generations may be haploid and their occasional fusion produces the sexual generation with double the number of chromosomes, which reducing produce haploid apomictical generations again. Thus we find all stages, from complete bisexuality to the unisexuality of virgin birth, down to no sexuality at all.

CHAPTER XVI

THE SMALLEST LIVING ORGANISMS

SO far we have been dealing with plants and animals that are made up of multitudes of cells, all of which are working together to form the complete organism. These include the higher plants and animals with which we are familiar, being visible to the naked eye. We will now for a moment glance at another vast world of living beings which are so minute that they live their lives unnoticed by the great majority of mankind and can only be observed and studied under the microscope. These are the great hosts of organisms which consist of a single cell or even less, since they may be uni-cellular with a complete or extended cell, or pre-cellular with an incomplete or primitive kind of cell. Some of these microscopical uni-cellular and pre-cellular organisms, which inhabit every corner of the earth, in land, water and air, can be distinguished by their habit of life as plant-like or animal-like, but in many other cases it is quite impossible to classify them in this way, since in some respects they may be plant-like while in other respects they are animal-like, while others again can be likened to both or neither. For this reason it has been found convenient to place many of these uni-cellular and pre-cellular organisms in a distinct kingdom called Protista.

Although these tiny one-celled creatures appear to be so simple, a glance at them through the microscope quickly shows us that some of them are extremely complex. So far we have been dealing only with one part of the contents of the cells—the chromosomes of the nuclei—but if we examine a cell we find that there are many other inclusions, perhaps not so interesting but at any rate vital to the life of the cell, and to the life of the chromosomes which bear the precious genes which carry the heritage of all the ages into the far distant future.

Fig. 178 shows a general diagram of a cell and its contents. Rather above the centre is the nucleus in a typically rounded form. Within it is the nucleolus, a round body about which there is much discussion but very little is known concerning its functions. Here we cannot see the individual chromosomes, since they are at rest, but we can see the reticulate threads into which they appear to turn during this stage. In animal cells there is usually a small body known as the central body or aster (fig. 179), which appears to take a directing part in the mechanism

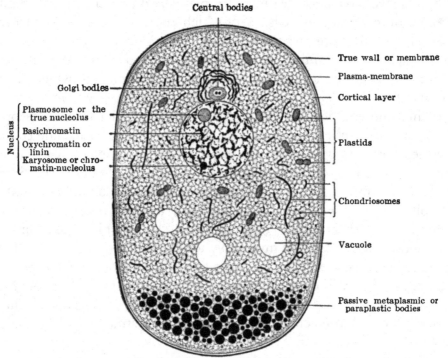

Fig. 178. A diagrammatic representation of a cell, showing the principal constituents. (After Wilson.)

of cell division but which is commonly absent or at least invisible in plants.

The material round the nucleus is usually recognised under the general term cytoplasm, but its components vary considerably in nature, origin and function. It is unnecessary to describe them all, and it will be sufficient to deal with the more striking ones. The whole of the cytoplasm is full of granular bodies, which are suspended in the clear ground substance which is more or less viscid. These vary much in size, shape and chemical composition. Many of them may be regarded as food supplies. The plastids are bodies chiefly characteristic of plants and are extremely important, since they have the power of producing vital products such as starch, pigments of various kinds, and possibly fats.

Perhaps the most striking inclusions of the cell are the chondriosomes, which are small granular or rod-shaped bodies which occur constantly

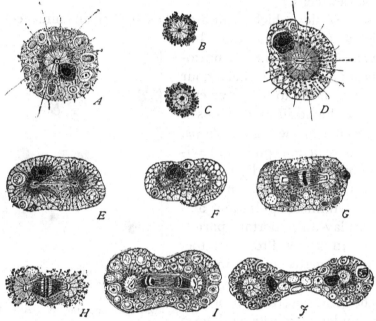

Fig. 179. Division of the one-celled heliozoön (Sun Animalcule) *Oxnerella*, showing the mechanism of its division, the appearance of the chromosomes and their equal partition into the two new individuals which will arise when the partition breaks down between the two new nuclei *J*. (After Dobell.)

in all cells whether in animals or plants. Much interest has been displayed concerning these in recent years, since it has been suggested that they may be the bearers of some part of the hereditary characteristics, but no proof of this is as yet forthcoming. It should be emphasised here that there is no evidence from genetics as to the inheritance of characters by any other means than by the chromosomes, except perhaps rarely by the plastids which appear in a few cases to be carrying free genes transmissible to the daughter cells. Since the genetical evidence is now so extensive it is safe to conclude that the gene is the sole basis for hereditary transmissions from the lowest organisms to the highest. In a living state the chondriosomes are hardly ever at rest, but wriggle and rapidly move about the cells from one position to another. They play a prominent part during the formation of the germ-cells and when the cell divides they are distributed in approximately equal numbers to the two new cells.

The Golgi bodies, which occur almost exclusively in animal cells, have given rise to much discussion. They are rather similar to the chondriosomes in appearance and behaviour but lie very near to the nucleus. Vacuoles are found in many kinds of cells, especially in the higher plants. They are more or less spherical cavities containing a watery liquid, and some forms of them in a living state show a rhythmical pulsation and appear to play an important part in excretion. In many Protozoa, too, solid food is digested in the interior of the vacuoles.

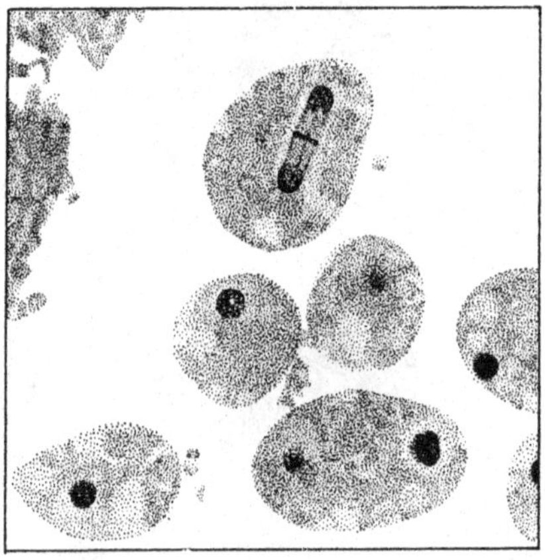

Fig. 180. Divisions in Amoeba (*Amoeba limax*). Upper cell dividing; lower cells, the two new individuals just breaking away from one another. (After Calkins.)

The chemical constitution of protoplasm (under which general term the whole contents of the cell are known) is very complex. It is made up largely of proteins and their various derivations, lipoids or fats, carbohydrates and inorganic salts, together with a large amount of water. There is a great similarity between animal and plant protoplasm, though relatively the proportions of its various constituents may vary considerably in different organisms and even within one species in different physiological states.

Having seen the complexity of a living cell we are not surprised that it is capable of sustaining life on its own account. Many of the minute uni-cellular forms of life indeed have not reached this state of complexity but others have attained an even greater. Fig. 181 illustrates a group of various kinds of uni-cellular Protists highly magnified, and a glance at the figure shows the difficulty of classifying these organisms. The most complex are usually placed in the sub-kingdom or phylum Protozoa, although some of the most simple forms of *Amoeba* are also

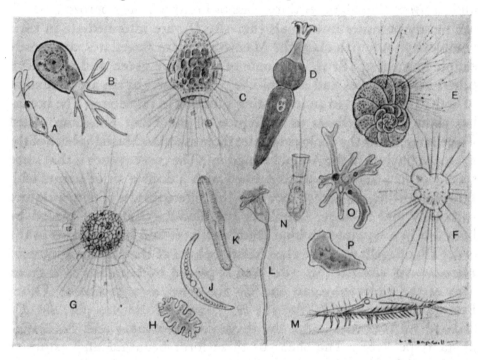

Fig. 181. Showing some of the more complex uni-cellular organisms. *A*, a Flagellate, *Trichomona*, from the mouth of man; *B*, *Cryptodifflugia*, an amoeba-like organism with a horny shell one two-hundredth of an inch long; *C*, a Radiolarian from the sea with a flinty skeleton; *D*, a Sporozoan, *Corycella*, an intestinal parasite of a waterbeetle; *E*, a Foraminiferan with a net of protoplasm; *F*, *G*, two kinds of Sun Animalcules with rays composed of living substance; *H*, *J*, two uni-cellular plants belonging to the Desmids; *K*, a Diatom, another plant; *L*, a Ciliate, *Vorticella*; *M*, another Ciliate, *Stylonychia*, crawling by means of bristles formed by joined cilia; *N*, Flagellate, with a collar; *O*, *P*, two types of *Amoeba*. (After Wells, Huxley and Wells, *Science of Life*.)

found in this group (figs. 180, 181, *O*, *P*). The Protozoa have been divided into four great classes, the most complex and elaborate of which are the Infusoria (Animalcules) and these are subdivided into two orders Ciliata (Ciliates) (fig. 181, *L*, *M*) and Heliozoa (Sun Animalcules) (figs. 179, 181 *F*, *G*). A second class Sarcodina (Rhizopods) contains the complex orders Foraminifera, with limestone shells (fig. 181 *E*), and Radiolaria, with flinty shells (fig. 181 *C*), and this class also includes the order Amoebina with the simplest forms of *Amoeba* (figs. 180, 181 *O*, *P*), and intermediate forms like *Cryptodifflugia* (fig. 181 *B*) and *Euglypha* (fig. 183). A third class of Protozoa, the Sporozoa, which are mostly parasites in animals (fig. 181 *D*), are intermediate in their complexity. A fourth class, the Mastigophora or flagellates, are mostly simple forms (fig. 181 *A*, *N*), some of which are green and plant-like, others are colourless and animal-like, while others again are a mixture of both. This has led to an interesting situation in taxonomy. The botanists claim the flagellates as a division of the Plant Kingdom called Flagellatae, while the zoologists claim them as a class Mastigophora of the Protozoa phylum of the Animal Kingdom. The consequence is that some species of flagellates are distinguished with a double set of names with different spellings in accordance with the different rules of nomenclature in botany and zoology. Similarly the dino-flagellates are placed by botanists in a separate division Dinoflagellatae and by zoologists in the order Dinoflagellata of the class Mastigophora of the phylum Protozoa. *Haematococcus pluvialis* (fig. 182) also is placed by botanists as a green alga of the Chlorophyceae and by zoologists as a Protozoa. Other Protista are the desmids (fig. 181 *H*, *J*) and the diatoms (fig. 181 *K*), claimed by botanists under the divisions Conjugatae and Bacillariophyta respectively. The shells of these form beautiful and exquisite pictures under the microscope.

Another primitive and generalised group distinct from any other is the Myxomycetes (Slime Fungi), also called Mycetozoa (Fungus Animals), placed by botanists as a separate division and by zoologists as an order of the Protozoa (Borradaile and Potts, 1932). Normally the species consist of a naked slimy mass of protoplasm (plasmodium), formless, without cells, but with thousands of nuclei, like a gigantic colony of amoebae, slowly creeping along in all the colours of the rainbow (bright yellow in the well-known "Flowers of Tan") and feeding on decaying matter. In adverse conditions it passes into a resting condition in the form of horny capsules containing about twenty nuclei and has been known to remain apparently lifeless for three years. Moisture brings it to life and movement again, when its peculiar methods of reproduction are manifest. Its fructification is like that of the fungi, producing stalked sporangia containing spores like toadstools and mushrooms. The minute ripe spores hatch out a microscopical organism

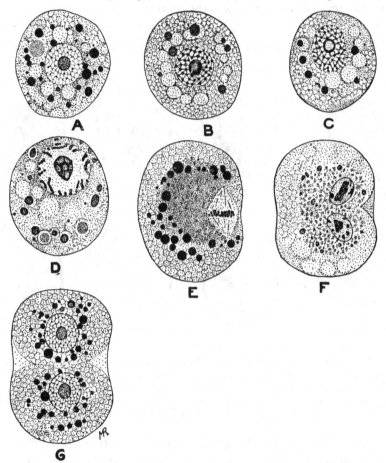

Fig. 182. The division of *Haematococcus pluvialis*. (After Minchin.)

combining the characters of an amoeba and a flagellate. These reproduce themselves by division, collect together, fuse sexually in pairs, making a small slimy mass of protoplasm. These masses collect together whenever they meet, ultimately forming a large colony, and thus completing the cycle of this generalised creature which combines the characters of an amoeba, a flagellate and a fungus and is a veritable object lesson in evolution. The sexual phases of these complex primitive organisms are well marked by a reduction of chromosome numbers, as in the higher plants and animals. The numbers of chromosomes vary in different classes, orders, families and genera.

The fungi constitute a large and distinct group of lowly organisms which has been subdivided into several classes and orders and more than

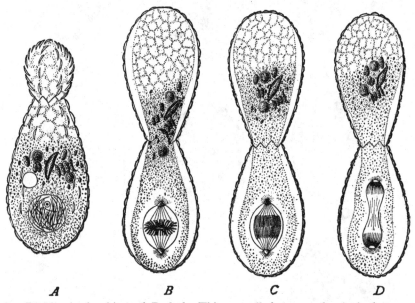

A   *B*   *C*   *D*

Fig. 183. Division in the rhizopod *Euglypha*. This one-celled creature has its body surrounded by a firm shell, so cannot divide in half in the normal fashion. It therefore divides by budding from the opening of the shell (*A*), the nucleus meanwhile dividing in the old individual. When the division is completed one of the two new nuclei will wander out into the bud which will then break off and form a separate individual. (After Wilson.)

100 families with multitudinous species. They are usually regarded as plants and in their structure resemble somewhat the green algae of the division Chlorophyceae, but they differ fundamentally from these since they have no chlorophyll and are not green, on the contrary they exude digestive juices like animals, and all are parasites living on the organic substances made by green plants or saprophytes living on the decay of living matter. Fungi reproduce themselves by spores which germinate and develop into fine white filaments or threads, which grow and inter-

lace into a soft felty mycelium. Each thread consists of an extensive uni-cellular continuous cylinder containing numerous nuclei. There is often a division into two or even four "sexes" in the sense that some threads will fuse together while others will not. Their fructification is peculiar and distinctive, being usually the most visible part of the organism, e.g. in the familiar toadstools and mushrooms. Altogether the fungi constitute a distinct kingdom of organisms.

As parasites fungi are very destructive and a plague to their plant hosts in the form of mildews, rusts and other diseases; they also cause ring-worm in man. In the form of edible mushrooms and indispensable yeasts, however, they are useful to man and the yeasts are among the smallest of the fungi (fig. 184). It is remarkable that notwithstanding the lowly organisation of the fungi their chromosome complex is as highly organised as in many of the higher plants and animals, though as a rule they are fewer in number, the gametic sets varying usually between two and eight.

The most simple forms of the Protista are the uni-cellular and pre-cellular blue-green algae (fig. 191) and the pre-cellular bacteria (figs. 185, 190). Formerly the blue-green algae were classified under a distinct division, Cyanophyceae, but more recently the blue-green algae and the bacteria have been placed as classes Schizophyceae and Schizomycetes, respectively, of the division Schizophyta. The rose-purple coloured species of bacteria of the family Rhodobacteriaceae form a connecting link between the bacteria proper and the blue-green algae, while the blue-green alga *Nostoc*, with its chains of globular pre-cells without nuclei, approaches the bacterian *Streptococcus* with similar chains of free pre-cells.

As a general rule the majority of these uni-cellular and pre-cellular organisms increase by cell division. When one grows too large to maintain itself in that condition the old cell divides in much the same way that the cells in our bodies divide, except that the two new cells formed do not remain joined in a common body but become separate individuals. In some of the higher species of uni-cellular animals and plants, colonies of cells are formed and often some form of sexual interchange takes place, in some cases in every other generation, one generation reducing its chromosome number, and fertilisation or a fusion of

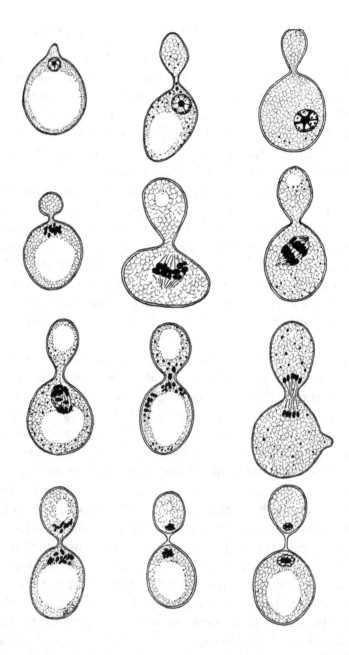

Fig. 184. Division in a Yeast Fungus (*Saccharomyces cerevisea*) by a process of budding, one of the two nuclei formed by the division of the chromosomes passing into the bud and thus forming a new complete cell. (After Kater.)

the cells taking place in the next. The ordinary cell division occurs with no reduction of the chromosome number.

In some of the most elementary species of blue-green algae and bacteria no chromosomes have yet been observed, in fact some extremely simple ones appear to have little cell differentiation at all but to be sub-cellular organisms too low in the scale of life to produce cell

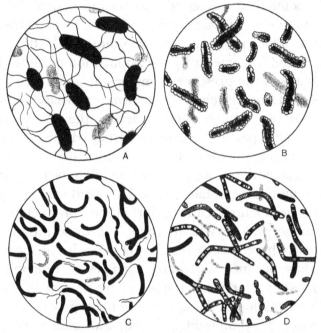

Fig. 185. Bacteria causing various diseases: *A*, typhoid fever, *Bacterium typhosum*; *B*, diphtheria, *Corynebacterium diphtheriae*; *C*, cholera, *Vibrio cholerae*; *D*, tuberculosis, *Mycobacterium tuberculosis*. (After Wells, Huxley and Wells, *Science of Life*.)

structures, though no doubt here the genes are present in a more or less free state and not held together in chromosomes, in the same way that the cells lie free in these lower organisms instead of being joined together. All the more highly organised uni-cellular species, however, have chromosomes and behave in just as orderly a fashion as the multi-cellular animals and plants, except that they exist singly instead of in joined groups of cells.

In bacteria (figs. 185, 190), although these minute sub-cellular microbes are for the most part without any visible differentiation in the

form of nuclei and chromosomes and without any visible sexual organisation, some form of heredity and mutation mechanism must be ad-
mitted, since various workers have succeeded in raising many different
strains in their cultures which breed true and behave in the same way
that mutations do in the higher organisms. Indeed the mutations in
bacteria are analogous in many ways to the mutations found in parthenogenetic and apomictical species of animals and plants, except that the
bacterial genes are presumably single and not in pairs owing to their
asexual origin. The fact that various mutational differences can be
artificially induced in bacteria by the application of ultra-violet rays is
highly significant of the presence of molecular genes (see p. 288). The
variant rough and smooth colonies of *Bacterium typhosum* have been shown
to be due to differences of chemical structure, the very virulent smooth
colonies possessing a particular protein-carbohydrate complex which is
missing in the rough colonies which have lost their virulence and the
mutations from smooth to rough breed true often for many generations.
Other marked mutations in bacteria which breed true are spore formation, pigment production, capsule formation, production of mucoid
substance, and so on (Arkwright, 1930). If these minute organisms,
which apparently have no aggregations of genes in the form of chromosomes, can thus produce hereditary changes of function, form and
colour, one must conclude that the genes are present in a more or less
free state. Confirmation of this is found in the reproduction mechanism
recently discovered by Stoughton (1929) and confirmed by Barnard
(1930), who find that at the time of fission the chromidia aggregate
together in the form of an ill-defined "nuclear" mass which divides into
two more or less equal parts and forms two new cell units. In some of
the higher species of bacteria the chromidia aggregate and form a spiral
or filamentous nucleus resembling a single chromosome (fig. 190). The
coming together of these scattered fragments to form chromosomes and
nuclei must have constituted a tremendous upward step in the pathway
of evolution, furnishing a reliable and permanent mechanism of heredity
and the foundation of the differentiation of sex.

More minute and more simple even than the bacteria are the viruses
or filter-passers.

Lately a considerable amount of work has been done with organisms so ultra-minute that they are capable of passing through the finest laboratory filters and collodian membranes. The germs of several of our most common diseases belong to this category, and to this fact is due the difficulty that has been experienced in isolating them and discovering means of preventing their attacks. It is evident that these minute microbes form another world of teeming life entirely beyond our ken and beyond the range of our most powerful microscopes, in the same way that our recent ancestors were unaware of the fact that the whole world swarms with bacteria and other lowly forms of life, some useful for our protection and others harmful for our destruction.

Owing to the fixed wave-length of visible light the smallest objects that can be seen distinctly, i.e. are definitely resolvable, under the most powerful microscopes are about a quarter of a micron in diameter ($250\ m\mu$), i.e. one four-thousandth of a millimetre or one hundred-thousandth of an inch. Ordinary bacteria are about one-half to three-quarters of a micron in diameter (500–$750\ m\mu$), so that they are in size not far from the limit of microscopical vision. Smaller even than bacteria are these ultra-microscopic organisms, which can be passed through a collodian filter, whose pores are less than a quarter of a micron in diameter. These are usually beyond microscopical resolvability, but when tested experimentally are found to produce various diseases in animals and plants, such as foot and mouth disease in cattle and swine, distemper in dogs, smallpox, typhus, yellow and trench fever, rabies, measles and influenza in man, mosaic and other diseases of the potato, tobacco and many other plants. They are generally known as Virus Diseases. So far it has only been found possible to detect those filter-passers which are definitely lethal or harmful to animals or plants, their presence being discovered by their disease-producing qualities. Judging from their nearest allies the bacteria, however, in which the harmful types are partly balanced by beneficent and useful types, many of which are a necessity to life, it is highly probable that there are filter-passers which are beneficent also, like the bacteriophage.

In 1925, a great step forward was taken when Barnard succeeded in securing images of what are presumed to be the virus organisms which cause pneumonia in cattle, by illuminating the microscope with the shorter waved ultra-violet light, which has no effect on the retina of the eye but which blackens a negative and produces a photograph. Fig. 186

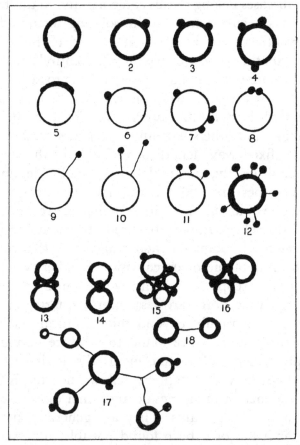

Fig. 186. Life beyond the microscope. The minute organisms which cause pneumonia in cattle made visible by the ultra-microscopic methods of J. E. Barnard. Tiny buds appear on the surface of the vesicles and often stay attached even after their growth to fully formed organisms, thus forming numerous clusters of varied shapes. (After J. E. Barnard.)

shows the result and introduces us to a new world of living microbes, smaller than the smallest bacteria.

These apparently living and reproductive particles which pass through the coarser filters are about one-fifth of a micron (200 $m\mu$) in diameter, afterwards they swell up into hollow spheres six or seven times as large, when they become visible under the microscope in ordinary light. Recent demonstrations by Ledingham and others (Gardner, 1931) of the tiny bodies visible in vaccine-lymph show that they are clearly of the same nature and order of size. The staining process seems to enlarge them and bring them just above the limit of resolvability of the microscope. Since they can be observed in the tissues after injection and appear to multiply and to be ingested by phagocytes just like cocci and other bacteria, there can be little doubt that they are microbes and they may well represent the actual virus agent. The large cell-inclusion bodies found in virus-infected tissues in man and other mammals, fishes, birds, insects, and plants of at least thirty-two families are essentially alike in structure, form and appearance; in animals they are usually known as "virus bodies" while in plants they are frequently distinguished as "X bodies". These bodies have been shown to consist of innumerable massed minute "elementary corpuscles"; thus if the cell inclusions in fowl-pox are crushed they appear to consist of minute bodies of regular size embedded in a hyaline substance, and there is a general consensus of opinion that these virus bodies are not gross organisms as was first believed, but are simply masses of virus particles each encased in a lipo-protein material arising from the reaction of the virus on the cytoplasmic contents of the cell. That these virus bodies should resemble one another so closely in the various cells of the Animal and Plant Kingdoms is surely significant of the natural relationship of the animal and plant viruses. In plants the detailed processes and mechanism of formation of these virus bodies have been worked out and filmed in living material of a tomato virus in *Solanum nodiflorum* by Henderson Smith and Sheffield at Rothamsted. The film shows clearly that the virus bodies are formed by the coming together in successive aggregations of minute granules usually converging towards the nucleus along the protoplasmic streams of the cytoplasm. Consecutive stages of the same processes of development of the virus bodies have also been clearly observed in fixed preparations of the mosaic *X* virus in the potato by the author, which fully confirm the particulate nature of the virus bodies (Salaman and Hurst, 1932). The author estimates the size of the granules observed to be of the order of 300 $m\mu$, but since the filter estimates of the size of the similar tobacco mosaic virus appear to be of the order of 30 $m\mu$ it is highly probable that the virus granules observed in the potato and tomato viruses are themselves aggregations of more minute particles which are unresolvable under the microscope with ordinary light. Further, in view of the estimated size of colloidal particles (in a fresh 1 per cent. haemoglobin solution), as of the order of 30 $m\mu$, it cannot be excluded that the virus organism or agent may be smaller still, since it might be adsorbed on a colloidal particle.

Barnard has recently been able to photograph smaller but similar organisms than those shown in fig. 186, and these are about a tenth of a micron (100 $m\mu$) across, from tumorous or cancerous growths of fowls and men, though whether these organisms are in any way a cause of cancer in man seems to be still an open question.

Finally, in the bacteriophage we reach the present limit of the smallest particles of matter that may be regarded as alive. It is estimated to be about one-fiftieth of a micron (20 $m\mu$) in diameter with a range of size from 8 $m\mu$ to 30 $m\mu$, and the question of its status has been a matter of lively controversy. This exceedingly minute filter-passer appears to break up and consume bacteria and by so doing may be beneficial to man. The fact that bacteriophages are able to increase their numbers rapidly and only consume living bacteria suggests that they are alive and of the same nature as a virus, although some authorities regard them as a non-living chemical ferment with the property of self-reproduction which acts only on living matter. Whichever may be the true interpretation, we seem here to have reached the borderland between the living and the non-living.

From the larger living microscopic bacteria and cocci measuring 750 $m\mu$ or more in diameter down to the smaller half-alive ultra-microscopic bacteriophages of 10 $m\mu$ or less there is a continuous series of sizes. There is no gap between the smaller bacteria and the larger virus organisms, nor between the smaller virus agents and the bacteriophages, neither in their measured sizes nor in their specific reactions with their hosts. Except in a few cases the ultra-microbes are too minute for their structure to be distinguished, yet notwithstanding this, the visible effects of their inherited specific actions provide evidence of the presence of specific genes or similar determinants, even in the most minute of the bacteriophages, and there can be little doubt that these genes are few in number and exist in a free state as in the bacteria. The faithful reproductions of distinct histological characters in the transmissible viruses of fowl tumours leave no doubt as to the presence of a specific virus gene in the defective cell. Each virus gene can be transmitted by artificial infection to a normal cell and the results are constant and specific in every histological detail, so that one gene will determine a spindle-sarcoma, another gene an ostreo-chondro-sarcoma and a third gene an endothelioma.

The permanent change of the human smallpox virus to that of the cowpox virus by passages through calves, whether it is due to mutation or to the dropping out of one of two viruses in inoculation as in Salaman's (1932) potatoes, in either case clearly involves the presence of specific genes or determinants. The X, Y and Z mosaic viruses and the R leaf-roll virus in the potato give definite and different specific reactions in various hosts, whether single or combined in a virus complex, and the experimentally demonstrated fact that the Y virus can be picked out of a complex of viruses by an insect *Aphis* (Kenneth Smith, 1931) provides evidence of the existence of genes or similar determinants in these viruses.

Salaman's remarkable study of the reactions of the X, Y and Z virus complexes in the potato show that the virus elements behave not as a mere summation or mixture, nor as a chemical valency, but as an organic linkage whose strength varies with the nature of the viruses and the internal environment of the plant host. In other words, the interactions and reactions of the elements of the virus complex are analogous to those of a complex of genes in the chromosomes of a cell nucleus.

When we come down to the extremely minute half-alive bacteriophage, some of which only measure 10 $m\mu$ across, i.e. one-hundredth of a micron or one hundred-thousandth of a millimetre, equivalent to the estimated size of a large molecule of albumin, we get into even deeper waters. There is, however, definite experimental evidence of the existence of pure races or strains of bacteriophages, each of which has a characteristic and specific range of action on diverse races of bacteria which is maintained after twenty passages of cultivation, lysis and filtration (d'Herelle, 1930).

Burnet (1930) enumerates six definite specific qualities found in pure races of bacteriophages, viz. (1) activity against other bacteria, (2) power to provoke qualitatively specific resistant strains, (3) resistance to heat, (4) power to induce specific antilysins on injection into animals, (5) size of their plaques, and (6) appearance of their plaques. Since these specific qualities are racial and constant, the most simple interpretation is that each of the six is organised by a specific gene. It is interesting to observe that although the bacteriophage is in general essentially lethal in its action, yet of the six genes it carries, three at least are definitely non-lethal in their action. Others may prefer the more complex interpretation that the determiners are chemical enzymes or catalysts specifically present in the bacteria themselves and this may be so, although it is only pushing back the problem a stage farther. Ultimately the specific enzymes or catalysts are in general either determined by genes or are themselves autoenzymes or autocatalysts which are equivalent to progenes or primitive genes. Indeed at this stage of life or half-life there is little to be gained by classing these molecules as living or non-living, since they are on the borderland of life and matter. It is highly probable that the half-alive protogenes and their immediate successors the progenes represent the earliest stages of the evolution of life from matter. Thus by the path of bacteriological experiments we are led to consider Muller's speculative suggestion that the bacteriophage may be actually a single free gene. Muller's suggestion was based mainly on a comparison of the estimated sizes of the bacteriophage and a gene of *Drosophila*, which at that time were reckoned to coincide more or less at about 20 to 30 $m\mu$. Since then other estimates have brought the *Drosophila* gene up to 60 and 77 $m\mu$ and the bacteriophage down to 10 and 8 $m\mu$, while Belling (see p. 46) estimates that the covered genes of the lily are much larger than those of the fruit fly, the split chromioles being of the order of 160 $m\mu$. It is natural to expect that genes differ in size even in the same species (cf. unilocal genes, p. 285), and it is questionable how far one can usefully compare the sizes of the fixed colonial genes of *Drosophila* with the free single genes of the bacteriophage. On the above hypothesis that the smallest bacteriophage carries a gene, the size of the gene of the bacteriophage would be something less than 8 $m\mu$ in diameter, and would lead one to infer that the free gene carried by the bacteriophage is a progene rather than a gene or that the ultra-microbe itself is a simple progene not far removed from the protogene and the origin of life from matter.

CHAPTER XVII

THE GENE AS THE UNIT OF LIFE

THE nineteenth-century idea that the cell with its protoplasm is the unit of life has now been superseded by modern research. The cytoplasm of the cell is controlled by the nucleus and the nuclear plasm or chromosomes carry the genes which reproduce themselves as living units and by concerted action organise development and reproduction. In view of the vital importance of the gene it is worth while to consider carefully and critically what we know about it. In so doing it is necessary in the first place to draw a clear distinction between positive experimental knowledge of the gene and speculative estimates based on experiments, of what it may be in nature, structure and size. The genes are too minute to be visible under the highest power microscope, but genetical experiments in the higher organisms have fully demonstrated their existence and position in the chromosomes, where they are located in linear order side by side, each gene occupying a certain locus on the chromosome relative to other genes and each gene giving a specific genetical reaction. Thousands of crossing-over experiments have determined these relative positions with repeated accuracy in the Fruit Fly (*Drosophila*) and in other species of animals and plants. This particular experimental method, however, only demonstrates the relative positions of the genes in the chromosomes, i.e. it gives us the relative position of a gene as a Euclidean point, with position but no magnitude. More recent experimental methods with transmutations, i.e. translocations or displacements of sections of chromosomes from and to one another and with deletions and duplications of sections of chromosomes, have not only confirmed the relative positions of the genes determined by the crossing-over method, but have given us in addition the relative lateral distances between the genes expressed in terms of lengths of chromosomes. Recently a further advance has been made by the Moscow geneticists in the use of this method in their studies of multiple allelomorphs by the application of X-rays to the Fruit Fly (*Drosophila*).

An allelomorph is one of a pair of genes located at corresponding loci in paired chromosomes. As successive mutations arise at one locus, a series of multiple allelomorphs or unilocal genes is formed. Thus if A be the original dominant gene, then recessive mutations a^1, a^2 and a^3 arising from it give rise to a series of four multiple allelomorphs of which A is dominant to a^1, a^2 and a^3; while a^1 is incompletely dominant to a^2 and a^3, and a^2 incompletely dominant to a^3. Thus all are unilocal and generally allelomorphic to one another. Since all occupy corresponding loci in a pair of chromosomes, not more than two, i.e. one in each chromosome, can be present in the same individual, except in the case of polyploids. In the case of "scute", a character in the fruit fly investigated by Serebrovsky and Dubinin (1930), its multiple allelomorphs, scute 1 to scute 7, have been experimentally shown to represent varying lengths of chromosome or step allelomorphs, the dominant normal scute representing a longer length of chromosome which includes the lesser lengths of the recessive multiple allelomorphs. The gene for "scute" determines the presence of bristles on various areas of the body, while the multiple recessive genes, scute 1 to scute 7, represent the absence of some or all of these bristles in these areas (fig. 187). It is interesting to find that the bringing together of the shorter lengths by crossing will, in suitable combinations, produce individuals which approach or reproduce the normal scute. This important development by the Russian geneticists is claimed by them to be a demonstration of the divisibility of the gene, but it seems to do much more than this, since it presents a new view of the nature of the gene which hitherto has been quite unsuspected. Thus dominance is identified with longer length of chromosome or a larger field of genic activity, and recessiveness is associated with shorter lengths or smaller fields of genic activity. On this view dominance is quantitative rather than qualitative as formerly supposed. These results have far-reaching effects, since they show that homologous or paired genes may differ in size, the larger genes being dominant to the smaller genes which are recessive. In this particular case it would seem that the smaller recessive genes have arisen from the larger dominant genes either by loss, mutations of parts of the gene or by the inactivity of the gene in certain lengths of the chromosome

causing a restriction of the genic field or sphere of influence. More
recent work by Agoi (1930) and others in America apparently confirm

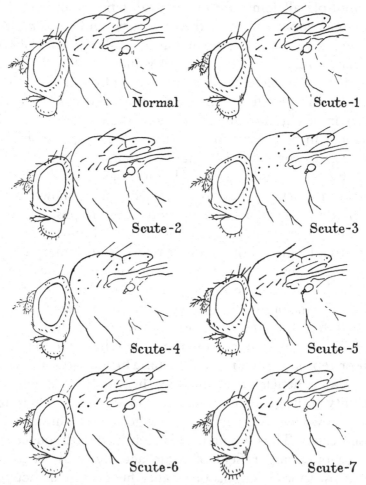

Fig. 187. Effects of the "Scute" genes in *Drosophila* (Fruit Fly). The unilocal multiple "scute"
allelomorphs affect different bristle groups in different degrees which are constant for each gene.
Some of them have effects on other characters than the bristles. The dots in the figure represent
absence of the bristles present in the normal "scute". (After Serebrovsky and Dubinin, *Journal
of Heredity*.)

these results experimentally with several other multiple unilocal genes,
but the precise interpretation of the results is still an open question,
which further experiments may answer. (Morgan and Bridges, 1931.)

TABLE 2

	Dorsocentrals	Supraalars	Verticals 1st	Notopleurals 2nd	Behind last pair legs	Verticals 2nd	Presuturals	Orbitals 3rd	Notopleurals 1st	Ocellars	Orbitals 1st and 2nd	Coxals	Postverticals	Abdominals	Scutellars	Humerals	Postalars
Scute 1	I	I	I	I	I	I	I	I	0	0	0	0	0	0	0	I	I
Scute 2	I	I	I	I	I	I	I	I	I	I	I	I	0	0	0	0	I
Scute 3	0	0	0	0	0	0	0	0	0	0	0	0	0	0	0	0	0
Scute 4	I	I	I	I	0	0	0	0	0	0	0	0	0	0	0	0	0
Scute 5	I	I	I	I	I	I	I	I	0	I	I	I	I	0	0	I	I
Scute 6	I	I	I	I	I	I	I	I	0	0	0	0	0	I	I	I	I
Scute 7	I	I	I	I	I	I	I	I	I	0	0	0	0	0	0	I	0
	a				b		c	d		e			f		g	h	i

Table showing the bristle-groups reduced (0) in different "scutes". (1) refers to bristle-groups unreduced. *a–i* represent centres for bristles along the chromosomes. (After Serebrovsky and Dubinin, *Journal of Heredity*.)

An analogous and probably parallel research by Julian Huxley and Ford on the genes of the Shrimp, *Gammarus chevreuxi*, carried out some years ago, takes on a new and important significance in the light of these later experiments. Huxley found that the eight shades of eye colours from black to pure red were represented by multiple genes, with black dominant to the other colours, and that the shades of the eye colours depended entirely on the speeds at which the black pigment was deposited, black being the most rapid and red the slowest, the rate of deposition depending on the particular gene concerned. In the light of the Russian experiments we might expect the dominant gene for black to be larger than the recessive gene for red, occupying a longer length of chromosome, and it is not unreasonable to suppose that a larger gene might be more rapid in action than a smaller gene.

These experimental researches indicate that genes have not only position but magnitude, and that their magnitudes are variable.

The establishment of the fact of the complexity of the gene, experimentally, leads us at once towards speculative concepts of its nature and structure. It is usually assumed that the gene has a physico-chemical structure and its size has been estimated in various ways, none of which is however quite satisfactory, since we do not know what outer coverings may enclose the gene itself. Assuming that the gene is in structure a chemical molecule or a group of molecules, the Russian experiments suggest that a dominant gene, e.g. "scute", may consist of a linear chain of equivalent molecules (each molecule probably with a similar or identical atomic constitution); individual losses, or inactivations of some of the molecules, giving the variable effects seen in the recessive genes, scute 1 to scute 7. We do not yet know whether all genes have multiple allelomorphs, although a large number of such are known, and many more genes may ultimately prove to be in this category. For instance, the large numbers of eye colours in the fruit fly may originally have been a complete series of multiple allelomorphs and the present scattered distribution of some of their genes in different chromosomes may be due to past displacements through translocations.

In such cases the highest dominant genes would be multi-molecular in constitution, while the lowest recessive genes might be uni-molecular in structure. Whether all quantitative mutations consist simply and solely of breaking up, or adding to, these chains of molecules (or their activations) we do not know. The question whether these "loss" mutations are due to actual losses of gene substance along the length of the chromosome, or whether they are due to inactivations of the gene in these areas, is at present difficult to determine. It may be that the normal dominant gene has a certain field or sphere of influence along the chromosome and that the recessive genes represent reduced genic fields of influence along the chromosome. The fact that both minus mutations (losses or inactivations) and the reverse plus mutations (additions or reactivations) are equally induced by the application of X-rays is highly significant (fig. 188), and points to the probability that these minus mutations represent inactivations in a genic field rather than actual losses of portions of the gene, and conversely that plus mutations in these cases represent reactivations in the genic field rather than actual additions of substance of the gene. On the other hand, deletions of sections of chromosomes involve actual losses of genes and duplications of sections of chromosomes involve actual additions of genes. In the same way non-disjunctions of genes, as of chromosomes, involve both additions and losses of genes. In the case of the large dominant "scute" gene the question arises whether it might not have been originally built up from the smaller recessive genes, scute 1 to scute 7, by the process of non-disjunction of genes in their divisions. Such a process would go far to account for the appearance of new dominant mutations with larger genes than those of the normal wild type.

If the difference between multiple allelomorph genes be mostly quantitative, depending on the number of molecules in the genes, it may be reasonably inferred that the differences between many non-allelomorphic genes are qualitative, in all probability involving molecules of different atomic constitutions. Here we get qualitative mutation in a molecule as distinct from quantitative mutation in a chain of molecules. Naturally qualitative mutations would be more complex in

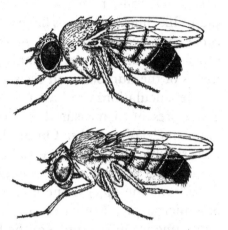

Fig. 188. A double mutation arising in *Drosophila* as the result of X-raying. Above, a fork-bristle fly with a normal eye, below, a normal bristle fly with "spectacled" eye which arose from it. The normal bristle represents a reverse mutation of the mutant gene forked back to the normal. (After Patterson and Muller.)

structure than quantitative mutations and probably much slower in action, since a qualitative change in a single molecule of a chain might have little effect until all or the majority of the molecules of the chain had become changed. These speculations, based on the Russian experiments, illustrate the extraordinary complexities possible in the structure of the gene.

With regard to the size of the gene, many attempts have been made to ascertain this, and several estimates have been made, especially with the genes of the Fruit Fly (*Drosophila*) and those of the Lily (*Lilium*). Different genera of plants and animals differ considerably in the size of their chromosomes, nuclei and cells, and it is to be expected that their genes will also differ in size. We have already seen from the Russian experiments with the fruit fly that multiple allelomorphs of the gene for "scute" appear to differ greatly in size among themselves and from the normal "scute" gene. The size of the plant or animal is, however, no criterion for the size of its chromosomes. Many of the larger trees have very small cells, nuclei and chromosomes, while comparatively small plants like lilies and beans have very large cells, nuclei and chromosomes. Since more work has been done on the genes of the fruit fly than in any other genus of plants or animals, there have naturally been more estimates of the numbers and sizes of its genes.

The calculations are based on the known and estimated numbers of genes present in measured and estimated lengths of the chromosomes of the nuclei of the cells. On this basis, in the fruit fly the largest possible gene size has been estimated to be of the order of 77 $m\mu$ in diameter and the smallest possible size 20 $m\mu$, giving an average gene size of 48·5 $m\mu$. A milli-micron ($m\mu$) is one-millionth of a millimetre or one-thousandth of a micron (μ). Muller estimates that the total number of genes in the four gametic chromosomes of the Fruit Fly (*Drosophila melanogaster*) must be not less than 1800, which represents 3600 in the four pairs of somatic chromosomes. This number, divided by the total length of chromatin, gives an average diameter of 50 $m\mu$ for the genes, a size approximately near that of some of the larger colloidal particles and some of the larger filter-passers. It is probable, however, that the genes are spaced to some extent and do not fill up the entire length of the chromosomes even with their outer coverings, in which case their actual size would be much smaller. Belling, in his interesting microscopical examination of the bivalent chromomeres of the Lily (*Lilium pardalinum*), estimates the total number of 2193 in a pachyphase cell with an upper limit of 2500. Since each chromomere consists of a set of two pairs of chromioles probably representing covered genes, the gametic chromosomes would contain not more than 2500 genes, while the somatic chromosomes would be carrying not more than 5000 genes. In the lily the chromosomes, and the genes, are apparently much larger than in the fruit fly, and the chromomeres visible under the highest power microscope (fig. 34) are estimated to measure approximately 670 $m\mu$ from centre to centre lengthways, and the space between them was usually more than half of this, so that the split chromioles might be estimated to measure 167 *m*μ in diameter, or more than twice as large as the estimate for the largest genes of the fruit fly. The actual gene itself would of course be considerably less than this even in the lily, and far beyond the

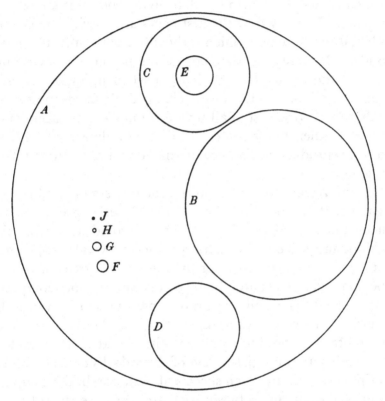

Fig. 189. A comparison of the sizes of particles and molecules, all magnified half-a-million diameters: *A*, particle just large enough to be seen through a powerful microscope; *B*, filter-passer; *C*, largest possible size of gene of *Drosophila*; *E*, smallest possible size of gene of *Drosophila*; *D*, bacteriophage; *F*, molecule of haemoglobin; *G*, smallest protein molecule; *H*, simple organic molecule, like chloroform, with five atoms; *J*, simple inorganic molecule of hydrogen gas. (After Wells, Huxley and Wells, *Science of Life*.)

visibility of the highest power microscope, which reveals objects no smaller than about 250 *m*μ. We can therefore hardly hope ever to see a naked gene under the microscope, although in the larger chromosomes of the lily and other liliaceous plants it is possible to see the chromomeres and even some of the chromioles in the pachyphase stage (fig. 34). In these plants, by the use of the ultra-microscope, it might be possible to record photographic images of the covered genes with greater distinctness.

That the protogene was the first manifestation of life on this planet is highly probable, since no form of life or living species is known to exist without genes. Precisely how the first genes arose, with the new power of growth and reproduction which is the all-important factor in life, we have no idea. Not only did they possess the power of growth but they also achieved the ability to keep it even when mutations so affected them that their whole reactions were changed. It seems most simple to suppose that they arose first of all by a fortunate combination of chains of carbon and other atoms present in that half-alive state midway between a self-reproducing chemical ferment and a living ultra-microscopic organism.

The first protogenes to appear were probably simple and free, living on inorganic matter alone, like some of the bacteria species to-day, and reproducing themselves rapidly. Mutations would arise from time to time, both quantitative and qualitative, their survival being controlled by natural selection. In course of time greater degrees of complexity would be attained and eventually, by co-operation, more complex units would be evolved. The building up of a protoplasmic covering by the genes was a great step in evolutionary progress. Those genes which, by fortunate mutations, produced the best-fitted and most suitable outer covering, would pass through the sieve of natural selection, and reproduce and multiply exceedingly, each successful mutation building up a more useful and potent barrier between itself and the outer world.

In the great kingdom of the Protista, in the classes of bacteria and blue-green algae, we find a relatively simple form of gene co-operation, where in the simpler forms the genes lie scattered in the protoplasm in the form of chromidia without a definite nucleus (fig. 190).

In other species an indefinite or definite nucleus appears containing scattered genes also in the form of chromidia (fig. 191). Later, in the higher and more complex species, a further and important evolutionary stage is reached by the linking of the genes together in the linear form of a chromosome (fig. 190) and afterwards into a number of chromosomes constituting a nucleus, as in the higher uni-cellular Protista and in the higher multi-cellular plants and animals, including man.

This evolution of the chromosomes bearing the genes in linear order provides a complex and consistent mechanism of heredity and reproduction which, together with the mechanism of mutation, transmutation and segregation of the genes and chromosomes and the evolution of sex, brings about the creation of new species, leading, under the control of natural selection, to Creative Evolution.

One of the most interesting and important problems in modern Biology is to discover the precise nature of the gene, how it is built up, how it grows and reproduces and how it controls and directs those chemical reactions which so profoundly influence the protoplasm, the cell, and the body in which it is contained. A wide field lies open here for the researches of the biophysicist and the biochemist. The experiments of geneticists have demonstrated that the addition or subtraction of genes has such a far-reaching influence on development, that the genes must be regarded as particles with specific properties which are capable of producing definite chemical reactions by means of which they control, direct and organise the processes of building up structures. These particles are presumably of molecular order, but possess the power to grow and reproduce themselves by utilising the matter around them. So far as one can see, this can only be achieved by catalytic synthesis and the genes may be regarded as autocatalysts. It is interesting to note that recent work on the behaviour of the nervous system shows that certain physiological effects are caused by the specific structure of the organic molecules involved. This is also being found for other bodily functions, the most striking being the activities of the various hormones. These show very varying effects, the influence of each one being dependent on its particular molecular structure.

So far, physiologists and colloid physicists, working at the problem of the origin of life, have taken for their basis the protoplasm of the cell. This has been chemically and physically tested and efforts have been made to reproduce it in the laboratories. Now, however, with the demonstration of the gene as the unit and basis of life, it is the study of the structure of these organic particles that will be the great work of the future, the protoplasm taking its proper place as a product of the basic living genes. The question whether life can ever be reproduced in the laboratory seems an extremely remote possibility now that the basis of life is found to be so much more minute and complex than had been anticipated. One might perhaps by a fortunate admixture of colloid particles reproduce protoplasm, but when it comes to reproducing complex living molecules, such as one believes the genes to be, it is altogether a different matter. One can hardly believe that it will ever be possible to bring together in the laboratory the necessary atoms to create a living gene, and even supposing this feat should be achieved there would still remain the task of bringing the genes together, of causing them to build up protoplasm, chromosomes, cells and all the complex structures which arise from their co-operative working. Unless this could be done their presence could only be detected, like that of the bacteriophages, by their action upon minute bodies which are themselves only just visible under the microscope, and this would not be a very satisfactory demonstration of created life.

Up to now the only work that has been done at all approaching the experimental study of the structure of these units has been the bombardment by X-rays, and, so far, this has only been used as a means of ascertaining the mutation effects of radiation on the genes, resulting in the striking demonstration of the random nature of mutation and evolution. So much has been discovered of the structure of the physical world by the bombardment of atoms with various rays, that it is not too much to expect that within the next few years, by the same means, great discoveries will be made concerning the structure of the gene, and the great problem of the ultimate foundation of life may eventually be solved. For the next great advance we must look to a new branch of biophysics built up on the genetical experiments of Morgan and Muller, by which the structure of the genes may be analysed and the problem solved as to the cause and nature of the differences between the complex molecules which form the various chemical compounds, and those which have attained the power of growth and reproduction and which have given rise to the immense complexity of life on this planet. On the inference that the gene is a complex molecule or group of molecules with peculiar autocatalytic powers, we must admit a continuity of life with matter. In the light of the New Physics matter is now regarded as electrical action and we have a sequence from the electrons and protons through the simple and complex atoms to the simple and complex molecules, some of these producing autocatalytic reactions culminating in the power of self-reproduction coupled with the capacity for mutation, thus giving rise to protogenes and genes, those ultimate units of life which form the foundation of the great evolutionary edifice.

CHAPTER XVIII

PROCESSES OF CREATIVE EVOLUTION

AFTER attempting to bring together the more salient facts and outstanding discoveries of the new branch of experimental biology known as Genetics, it may be useful to review briefly the material dealt with and to draw such conclusions as may be warranted by the experimental evidence produced. So far as we know at present, the gene is the basic and ultimate unit of life which exists in all the species of living organisms from the simple microbe up to complex man. The old nineteenth-century units of life, the cell and protoplasm, are in the twentieth century regarded as complex products of more simple units, the living genes. The genetical discoveries of the last twenty years have demonstrated experimentally that in the higher plants and animals these genes are located in a linear series along the threads of the chromosomes of the nucleus of each cell, and are in their reactions to the environment and to one another the organisers of the characters of plants and animals (including man), causing and determining their growth and development, and are also the sole means of carrying forward these characters to future generations through the medium of the germ-cells.

The smallest and most simple living things known to us, smaller even than the microscopic bacteria, are the ultra-microscopic viruses or filter-passers, which are as yet little known but which some day may help us to solve the pressing problems of the structure, organisation and evolution of the genes, of their powers of growth and reproduction and how they produce the chemical reactions which in combination build up and maintain the complex living bodies of the higher plants and animals. The question of the status of the bacteriophage, the smallest filter-passer known, which may also be a virus and which attacks and breaks up living bacteria, has been a matter of considerable discussion, some authorities regarding them as living organisms (viruses), while others consider them to be bacterial enzymes with powers of self-reproduction. Whatever they are they appear to be on the borderland of the living and the non-living. Experiments demonstrate the existence of at least six distinct constant races of bacteriophages which involve the presence in each of a distinct gene or determiner of a similar nature. Since the smallest bacteriophage is of molecular order of size (8 $m\mu$) and is considerably smaller than the smallest estimated size of a gene of *Drosophila* (20 $m\mu$), it is concluded that the gene of the bacteriophage is very small and may be of the nature of a primitive gene (progene) or autocatalyst not very far removed from the original half-alive protogene which constituted the first step in the evolution of life from matter. So far there is no experimental evidence that the bacteriophage carries more than one gene, and its monogenic nature may be held to support Muller's speculative suggestion that the phage particle itself is a free gene. Some of the larger bacteriophages correspond in order of size with a colloidal particle of 30 $m\mu$, but it is not altogether excluded that the actual bacteriophage may be much smaller and carried by adsorption on a colloidal particle. The mosaic virus of the tobacco plant is estimated to be of the same order of size as the largest bacteriophage (30 $m\mu$), while the similar mosaic viruses of the potato (X, Y and Z) appear to be monogenic and may also be progenic in their nature. Other viruses in animals and man are of a larger order of size, varying from 75 $m\mu$ to 250 $m\mu$. Some of these may be polygenic like the bacteria but the tumorous viruses of fowls seem to be definitely monogenic. The various intra-cellular inclusion bodies found in the virus-infected tissues of various plants and animals (including man) have been shown to be composed of minute particles which aggregate in the cytoplasmic streams to form the virus bodies. Recent demonstrations show that some of these virus particles are definite organisms which are large enough to be resolvable under the microscope with ordinary light.

The larger animal viruses and filter-passers merge in size with the smaller bacteria and produce similar reactions on their hosts. There can be little doubt that the bacteria and cocci have evolved from the more simple viruses, since they represent an advance in size and complexity on the same lines and in the same direction. In order of size the bacteria range from 100 $m\mu$ to 1000 $m\mu$, the average width of the ordinary bacteria and cocci varying from 500 $m\mu$ to 750 $m\mu$. In the ordinary species of these sub-cellular microbes there is no visible nucleus, often no distinct cell wall, and there are no definite chromosomes or traces of sex fusion save in a few of the more complex species which are exceptional. That the bacteria carry genes has been experimentally demonstrated by the raising of mutational and constant genetical races or strains in many different species. Quite recently the hereditary mechanism of bacteria has been independently discovered by Stoughton, and Barnard, who have shown that at the time of fission the scattered chromidia come together in the form of an ill-defined nucleus which divides during fission and is clearly a part of the process. There can now be no doubt that the bacterial genes are located in the scattered chromidia as free genes and that these chromidia in bacteria are homologous with the linear chromomeres found in the higher organisms. This condition of grouped chromidia is also found in the simpler species of blue-green algae (fig. 191), flagellates, rhizopods and Protozoa. In some of the higher species of bacteria the scattered groups of chromidia evolve into a spiral filamentary nucleus resembling a chromosome, which divides into two at the time of division as in the higher plants and animals (fig. 190).

The coming together of these scattered groups of chromidia to form linear chromosomes represents a tremendous uplift in the mechanism of creative evolution. The development of a definite cytoplasm with or without a cell wall, giving rise to a cell structure and to uni-cellular organisms, was another great step, leading as it does to the multi-cellular higher plants and animals, including man. In the higher forms of the blue-green algae the chromidia become centralised as in *Chroococcus turgidus* (fig. 191 *A*) and eventually form a definite nucleus as in *Chroococcus macrococcus* (fig. 191 *D*), but without the formation of chromosomes, and during the division the nucleus divides. In *Glaeocapsa oeruginosa* (fig. 191 *B*) an irregular spiremoid nucleus is formed, while in *Merismopedia* (fig. 191 *C*) *Oscillatoria* and *Lyngbya* are found closer approaches to the chromosomal conditions of the higher plants and animals.

It is of course possible that the simple blue-green algae, bacteria and viruses have originated in the reverse way, and that the scattered groups of chromidia have devolved from normal chromosomes by a loss of the mechanism and consequent disorganisation and fragmentation leading to degeneration but not to death. Since the general trend of evolution is, however, on the whole creative and progressive, it seems more probable that these minute organisms represent an evolutionary stage of development in plants and animals. From these minute and simple viruses, sub-cellular bacteria and blue-green algae, we find a

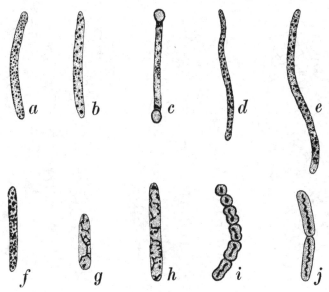

Fig. 190. *a–c*, bacilli of *flexilis* type with scattered chromidia (*c*, showing spore formation); *d–f*, similar type of bacilli from different host; *g, h, Bacillus saccobranchi*, chromidia coming together; *i*, chain of cocci with massive nuclei resembling a chromosome, some dividing; *j, spirogyra* type of bacillus with spiral filamentary nucleus. (After Dobell.)

gradual increase in complexity through the various types of one-celled protists (fig. 181), until in some of the higher forms of these we find a high degree of complexity comparable to that found in the higher multi-cellular plants and animals.

The formation of social colonies of single cells may have been the next great step in evolution (fig. 192). At first extremely simple, each cell living for itself much as if it were still free, the gradual growth of these colonies and consequently the necessity of co-operation in the matter of food supplies, the exudation of waste products and so forth, gave rise gradually to the specialisation of different groups of tissue cells until in the higher animals we find it in its most perfect condition, each part of the body having its own particular function to perform, being dependent on other parts for the performance of other functions. In these higher organisms the whole body serves only as a temporary protection and food supply for the all-important germ-cells which alone survive mortality and carry the race on into the far distant future.

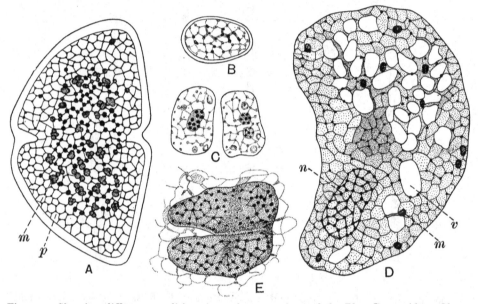

Fig. 191. Showing different conditions in various members of the Blue-Green Algae Chroo-coccaceae. *A, Chroococcus turgidus,* with scattered metachromatic granules (*m*), division just commencing; *B, Glaeocapsa* in which central region often shows a spireme-like condition; *C, Merismopedia elegans* with definite nucleus formed by an accumulation of chromidial material; *D, Chroococcus macrococcus* with a higher type of nucleus having a reticulum with chromatic granules; *E,* the dividing nucleus of the same. (After Acton.)

It is an open question, however, whether these first cell aggregates arose by the failure of individual cells to disunite or whether they arose by the breaking up of one large multi-nucleate cell into a mass of uni-nucleate cells. At the present time there are many of the simpler organisms which show no cell formation but contain a large number of free nuclei and show a certain amount of structure, e.g. in the case of the Green Alga *Caulerpa* which has "stems", "roots" and "leaves", if these may be so called, but no cell structure, the whole forming one large multi-nucleate cell. In the higher plants we get an example of free nuclei in the endosperm which forms round the young embryo as food storage and remains in this state for some time, finally, however, dividing up into uni-nucleate cells. The male and female gametes also form multi-

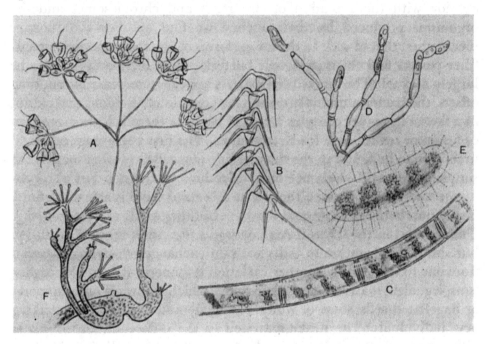

Fig. 192. Co-operation of cells. *A*, colony of collared Flagellates (see fig. 181*N*); *B*, marine Flagellate *Ceratium* in which individuals (which multiply by fission) may form chains temporarily; *C*, part of the filament of the Thread Alga *Spirogyra*; *D*, *Chlorodendron*, an organism in which the cells sometimes break away forming a link between uni-cellular organisms and the thread alga; *E*, *Collozoum*, a floating colony of Radiolarians; *F*, part of a colony of *Dendrosoma*. (Wells, Huxley and Wells, *Science of Life*.)

nucleate cells at some stages of their formation. The young embryo itself forms cells from the beginning in both plants and animals, and if there is anything in the recapitulation theory this would suggest that our remote ancestors arose more probably from an aggregation of cells than from the breaking up of multi-nucleate organisms, or it may be that there have been two independent modes of cell evolution. The lack of cell formation must in any case be a serious handicap, since no solidity of structure can be achieved without it, and to build up larger and stronger organisms the division of the protoplasm into cells becomes an urgent necessity.

As the complexity of the organism increases we find a gradual increase in the complexity of the gene and chromosome mechanism. Side by side with the evolution of the genes and chromosomes and the organisms produced by their agency we find another evolutionary mechanism, that of sex. In the lowest forms of life, the ultra-microscopic filter-passers and the microscopic bacteria, reproduction appears to be largely asexual. The larger filter-passers appear to increase by budding offsets, the bacteria multiply usually by constricted divisions and except in a few of the more complex forms of bacteria there is little suggestion of fusion or sex in these lowly organisms. The first beginnings of sexual union are to be found in the microscopic one-celled protists such as the simple flagellates, which not only reproduce by division but also conjugate and fuse in pairs, although at this stage there is still no definite distinction between male and female. In other kinds of Protozoa, e.g. Ciliates, and in the Thread Alga *Spirogyra* (fig. 192), there is no fusion but simply conjugation in pairs with an exchange of nuclear material from one to the other. Another variation is found in the simpler organisms, e.g. algal seaweeds and slime fungi, which throw off minute spores or flagellated cells, some of which conjugate and fuse together, making new individuals. The next great step in the differentiation of sex in plants and animals is found in the formation of two distinct kinds of reproductive cells or gametes, the one larger, more passive and female, and the other smaller, more active and male.

As we proceed up the scale to the higher and more complex plants and animals we find this differentiation of sex into male and female becomes more pronounced until we reach the higher mammals, where sexual reproduction is apparently the sole means of reproduction, whereas in insects, edentate mammals and the higher plants sexual and asexual reproduction often exist side by side. In most plants, snails and some other animals we find hermaphrodites in which the same individual may function both as a male and as a female, while in oysters we have another form of sexual differentiation in which individuals may function alternately as males and females at different periods of their life history. In the various phyla of plants and animals we find all degrees of reproduction between sexuality and asexuality and there can be no doubt that the evolution of sex, side by side with the evolution of plants and animals, by the introduction, recombination and perpetuation of new characters, has provided a mechanism of syngamy and segregation which has been a most potent factor in evolution.

Genetical research in numerous families of plants and animals shows clearly that the evolution and origin of new species from old ones has followed different courses in different families and genera. The mechanism has been the same throughout, the mutations of the genes and the transmutations of the chromosomes, distributed by means of the sex mechanism, produce the changes in the characters and also the sterility necessary to form new species, but the type of mutation varies considerably in different families, some families showing gene or chromosome mutability in one direction and some in another.

The word "mechanism" applied to living organisms has always aroused a storm of protest in many quarters, but it is extremely difficult to find another word which so exactly describes the orderliness and organisation which we find throughout nature. If there was no kind of mechanism such as we find in the orderly behaviour of the chromosomes and if each cell behaved in a manner peculiar to itself instead of conforming to the rules of cell division, we should have nothing but monstrosities, without form or function. In fact existence would be a mockery and there would be no life at all as we know it. An illustration of this may be found in cancer cells, which have been shown by microscopical examination to be cells in which all regularities of cell and chromosome divisions have ceased (fig. 193). They go on dividing in any haphazard way, the chromosomes going into unequal groups, splitting and fragmenting or getting lost altogether so that there is no differentiation of the growing mass, no organisation and no control, with disastrous results to the unfortunate individual attacked by this terrible scourge. The cancer infection, when it attacks a body, apparently has the power to stop the regular mechanism of the cells and we get an illustration of what would happen if there were no orderly mechanism in nature. So that when we apply the word "mechanism" to evolution we do not mean that evolution has behaved like a machine but that it has followed certain determinate sequences which regulate it and lead it into the various channels in which it now runs, in spite of its fundamentally indeterminate and unpredictable nature.

The living mechanism of creative evolution in plants and animals lies in the genes and chromosomes, which not only organise and control

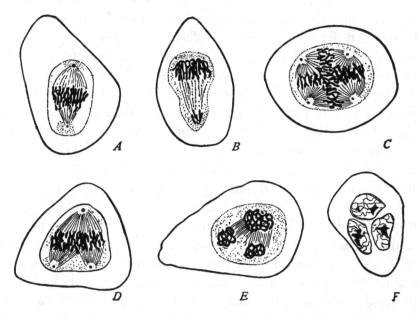

Fig. 193. Irregular chromosome and cell divisions in human cancer cells, resulting in unhealthy and abnormal cells lacking in differentiation. (After Galeotti.)

the destinies of each cell and of each individual but which, in the germ-cells, provide the possibilities and potentialities of each succeeding generation. By their orderly behaviour and the chemical reactions produced by them are built up the bodies of all living things, and their occasional mutations and transmutations provide the changes which have gradually given rise to all the diverse species in existence to-day. These changes are demonstrated by the experimental examination of the chromosomes and characters of the species of many different families and genera, and a comparison of their differences provides an interesting study of the multifarious modes by which creative evolution has proceeded. Some families and genera have species with chromosome sets which are apparently similar in number, shape and size but which carry different genes. Others have species with identical chromosome numbers but the chromosomes are variable in shape and size, while others again have chromosomes which are variable in number, shape and size. Some genera have species which are distinguished from one another by various constrictions in the chromatin forming segmented chromosomes and in some cases with minute attachments known as satellites or trabants at the end of one or more pairs of chromosomes. Many families and genera of plants have species with multiple sets of chromosomes (polyploids) which present a range of numbers which are multiples of a primary basic number. It is evident that all these various chromosome complexes have had different modes of origin and have arisen through mutations and transmutations. Those species with chromosomes more or less similar in number, form and size have become differentiated by means of changes in the gene complex through mutations or equal translocations of sections of chromosomes. These are in most cases primeval types of chromosomes that have persisted and remained constant, except in their gene complex, through long periods of secular time. Those species with chromosomes which are variable in number, shape and size have arisen by transmutations in which are included fragmentations, deletions, duplications, displacements, fusions or unequal translocations of chromosomes; by non-disjunction (additions), or more rarely by losses of sections of chromosomes, whole chromosomes or sets of chromosomes. Some polyploid species with multiple specific sets of chromosomes have arisen by hybridisation followed by a duplication of the gametic sets which has made the hybrids fertile. In plants this seems to be quite a common method of species formation, since during the last few years many new polyploid species and several new polyploid genera have appeared in genetical cultures in various genera in Europe and America. These new polyploid species and genera combine some of the characters of the original parent species and are quite distinct from either parent, they are completely intra-fertile and in most cases are sterile with the original parent species and each has a distinct and definite chromosome complex. In all respects, therefore, these new experimental species may be said to fulfil the requirements of a good taxonomic species, and although they arose in experimental cultures it is difficult to deny them the status of a species. Other polyploid species have arisen from mutations, transmutations and crossings in polyploid varieties, and hybridisation has been responsible for other modes of species formation, as in the *Caninae* and in the regular polyploid species of *Rosa*.

All these various transmutations which bring about changes in the chromosome complexes of varieties, species and genera have, together with gene mutations, led to the formation of new species and genera, and thus provide a mechanism of creative evolution. One of the most important results that have followed these findings is the discovery that the chromosomes in each species are remarkably constant in number, size and shape (except in their polyploid varieties which are incipient species). This discovery is vitally important to the systematist and taxonomist, since it means that the specific rank of any individual can now be definitely established by means of a critical and experimental examination of its chromosomes and characters, and the time is not far distant when these experiments will be regarded as the only true criteria of the systematic position of a plant or animal and a new exact science of taxonomy established on an experimental basis. Comparative studies of the chromosomes and characters of a genus, of the genera of a tribe and of the tribes of a family, will help to elucidate the problems of phylogeny and evolution. Much work remains to be done, but even now the work already accomplished has enabled us to present more definite ideas of the nature of a species and has made it possible to present a much clearer definition of the concept of a species than was possible in the days of Darwin (see p. 66).

The various processes and mechanisms in action in the formation of new species by transmutations of the chromosome complex are all of vital importance, but equally important are gene mutations, since these provide the foundation of the origin of new species, though not the sole and immediate cause of speciation. In their heredity, mutations follow

Mendel's laws and, when homozygous in later generations, represent the hereditary variations in all directions postulated by Darwin in his theory of Natural Selection, now firmly established as a law of nature. A single homozygous gene mutation gives rise to a new systematic form which, if it survives the rigours and trials of natural selection, becomes isolated and remains homozygous, may in combination with other gene mutations in the course of time develop into a variety, a sub-species, and ultimately, after a long period of time, into a species. This is an extremely slow process of speciation, but given long periods of geological time it is possible for original formal or varietal mutations ultimately to represent not only specific characters but in secular time to become sectional, generic or even higher in rank.

Certain characters are recognised as varietal in one species, genus or family, while in others the same or a similar character is recognised as sub-specific, specific, sectional or even generic. So far as one can see, the only distinction between these group characters is the differential duration of their stability in a state of nature. Formal and varietal characters are mutable, Mendelian, and relatively evanescent compared with specific characters which are less mutable and more stable and lasting in nature. In the same way specific characters in secular time are relatively fleeting compared with generic characters and with the higher characters peculiar to families, orders, classes and phyla, which the evidence from fossils conclusively shows to be more stable and durable than the lower groups. The nature of this differential stability in the higher and lower natural groups is not yet fully known, but genetical evidence shows clearly that the apparent stability of the higher groups does not hold good in cultivation and domestication to the extent that it seems to do in nature, since artificial selection and cross-breeding by man has in a relatively short time completely changed specific and even generic characters, as witnessed in the artificial races of dogs, horses, pigeons, poultry, roses, begonias, orchids and other garden plants and domesticated animals. In view of this evidence one can hardly escape the conclusion that the stability of the higher groups in nature is due to the action of natural selection, and that the higher taxonomic and morphological groups are in some way correlated or definitely associated with physiological functions which determine survival in nature, while the less stable varietal and formal groups are of less importance to survival in a state of nature.

A fact of considerable importance that has come to light in recent genetical experiments (Blakeslee, 1927; Hurst, 1929) is that when a specific, sectional, generic or family character does mutate the new mutation is at once varietal and Mendelian, it loses its high status and becomes simply a varietal character. For instance, in *Rosa*, the climbing habit of growth with its associated characters of flexuous branches and branchlets, hooked upper prickles and summer-flowering habit, is a sectional character of the old *Synstylae* of De Candolle and a specific character under the new genetical classification (Hurst, 1928). Under centuries of cultivation in China and Europe and also rarely in a wild state in China, this climbing specific character has mutated to a series of dwarf varietal characters with less flexuous branchlets, straighter prickles and perpetual-flowering habit which, genetically, behave as Mendelian recessives to the climbing specific character which now becomes a dominant varietal character. Similarly in *Datura* a generic and family character has mutated to a varietal character (fig. 135). Two important results arise from these experiments. First, they demonstrate that the higher specific, sectional, generic and family characters are represented in the chromosomes of the nucleus by genes in the same way that varieties are represented. Second, they show that varietal, specific, sectional, generic and family genes and characters do not differ from one another in kind but only in degrees of stability or mutability in nature which are apparent only under the influence of natural selection. Under cultivation and domestication the higher groups tend to be as mutable and unstable as the lower groups.

There can be no doubt that the characters which distinguish the higher groups of to-day first originated in geological time as formal or varietal gene mutations, but the process of the formation of species, genera and families involves much more than the simple fact of mutation. A species is made up of many homozygous dominant specific characters, some diagnostic and others non-diagnostic; consequently many gene mutations are necessary to make a new species and all these have to pass through the sieve of natural selection, amid the variable and often extreme changes of climate in different geological periods.

Moreover, another important and necessary quality has to be acquired by a variety or a sub-species before it can attain to the full rank and independence of a genetical species, and that is some degree of sterility when crossed with the parent species or with other species. Recent experiments show that this inter-specific sterility is acquired largely by transmutations, i.e. changes in the chromosome complex. Fragmentations, fusions, deletions, duplications, translocations, displacements, non-disjunctions involving additions or losses of genes and chromosomes, also additions or losses of whole chromosome sets, all lead to different groupings and balancings of the genes causing variation of characters and different degrees of sterility in the formation of the germ-cells in the first or second generation, leading in cases of extreme differences between chromosome complexes to complete inter-specific infertility.

It will be observed that the mechanism of creative evolution by mutations of genes and transmutations of chromosomes necessarily involves both gains and losses of genes in the evolutionary process. Gains of genes have been brought about by hybridisations, duplications of sets of chromosomes, non-disjunction of single chromosomes and by duplications of sections of chromosomes following translocations and displacements, all of which have been experimentally demonstrated. Losses of genes have been brought about by losses of sets of chromosomes, losses of single chromosomes by non-disjunction, and by deletions of sections of chromosomes following translocations and displacements, all of which have been demonstrated by experiment. Except in polyploids, however, loss of single chromosomes and deletions of sections of chromosomes have proved to be mostly lethal and hence of little evolutionary significance. Losses of whole sets of chromosomes have, however, proved to be viable and significant in evolution in certain polyploid genera.

That some diploid and lower polyploid species in nature, e.g. in *Rosa*, have been evolved from the higher polyploid species seems probable from the evidence of losses of whole sets of chromosomes observed under the changed conditions of cultivation during several centuries (Hurst,

1925, 1926). From the geological and genetical evidence it is highly probable that the polyploid species of *Rosa* were originally built up from older diploid species, either by hybridisation followed by the doubling of the chromosomes or by mutations and crossings in polyploid varieties or by both processes. It may be that in polyploid genera like *Rosa* and *Rubus* a rhythmic up and down evolution of species has occurred since the Pliocene Period, reaching its climax in the great Mindel Ice Age of the Pleistocene when the highest polyploid species may have reached the decaploid state. On the retreat of the ice sheets in Europe and America it is probable that some of the lower polyploid and diploid species emerged in a changed condition from the higher polyploid species by losses of complete sets of chromosomes. The few ancient monotypic diploid species of the tribe Roseae, now present only in the southern areas, are probably relics of the Pliocene and Miocene diploids driven south by the Ice Age and are now considered to represent distinct genera (Hurst, 1928). Losses of genomes in the offspring of polyploids have been reported by Darlington in *Prunus* and by Gates in *Oenothera*. The question of its application to other genera of plants and to animals requires further investigation, but it may be that subtractions or losses of chromosomes and chromosome sets are of importance in evolution as well as additions of chromosomes and chromosome sets. Since each chromosome carries large numbers of genes it is evident that the addition or loss of a section of a chromosome, a whole chromosome or a set of chromosomes, represents much greater changes in evolution than the mere change in single genes. For that reason chromosomal changes are here distinguished as Transmutations, while genic changes are called Mutations.

If the findings in *Rosa* are confirmed in other plants and animals we have here a new and unexpected evolutionary process of considerable importance, since, in addition to the ordinary creative process of integration from the simple to the complex, we have also the reverse process of disintegration from the complex to the simple, not as a degenerative process but as a normal process of the emergence of new species from old species. Experiments show that the simple diploid species are alone able to give complete expression to all their characters, each being specialised and adapted to occupy a peculiar niche of its own in nature, whereas the complex polyploid species is more generalised in its characters and habitats, being able to adapt itself and survive under adverse conditions where the specialised diploid species would perish. It may not be without significance that throughout the ages in every phylum of plants and animals it has usually been the lowly generalised species that have given rise to the higher specialised species. In this way evolution would be an alternating process, truly creative from the simple to the complex and truly emergent from the complex to the simple. The emergent diploid and lower polyploid species of the later Pleistocene (e.g. *Rosa*) would no doubt differ in some characters from the original diploid and lower polyploid species of the Pliocene from which the higher polyploids of the Pleistocene were built up. It is important, however, to emphasise the fact that all generalised species are not necessarily polyploids, since in the tribe Roseae there are in existence to-day five genera, probably relics of the Miocene roses, which are all generalised diploids with no polyploid representatives so far as known.

The experimental inducement of mutations and transmutations by artificial means is a most important work, since it gives us a clue to the probable causes of mutations and transmutations in nature. The irradiation of the Fruit Fly (*Drosophila*) by Muller and others, of *Nicotiana* by Goodspeed and of other plants and animals by other workers have produced the same random mutations and transmutations that have appeared from time to time in experimental cultures, the only difference being a rapid increase in the frequency of the mutations and transmutations after irradiation as compared with normal control cultures. In *Drosophila* this is estimated to be an increase of 15,000 per cent. This is a clear case of random or indeterminate mutations and transmutations caused by the bombardment of the genes and chromosomes by X-rays. It is especially interesting to observe that in every case so far investigated, the type of gene mutation or chromosome transmutation produced by artificial means is the same as that found occurring naturally under normal conditions, thus further demonstrating that each family and genus has its own peculiar type of gene and chromosome evolution.

De Mol in Holland, in a long series of experiments with Hyacinths, Narcissi and Tulips, found that they produced diploid and tetraploid (duplicated) pollen grains when the bulbs were submitted to extremes of temperature and dryness at the time of gamete formation as well as when irradiated with X-rays. Harrison and Garrett, working with Moths, apparently induced Mendelian melanic forms by feeding the caterpillars on chemically treated foliage, and further experiments on these lines are awaited with interest. It is possible that other extreme treatments with various substances and different kinds of radiations may demonstrate other external causes of mutations of the genes and transmutations of the chromosomes. Certain experiments with different temperatures apparently raise the incidence of mutations slightly, but the effect seems too small to be taken into serious consideration. Since the genes are trebly protected by the cell membrane, the nuclear membrane and the chromosome structure, they are well out of reach of ordinary external influences except through the food supply, and in any case the chance of these organic molecules being affected except under exceptional circumstances is rather remote. So far as the effect of irradiation is concerned, however, they are not so impervious, since the stream of electrons can pass through the atomic spaces of the body and cell structure, maybe hitting an electron here and there as they pass, which in its turn will create further atomic and molecular disturbances. Occasionally one of these chance hits occurring in a genic molecule will produce a permanent mutation, provided that it is not detrimental to the development of the organism and can survive the test of natural selection in times of sudden and extreme changes in the conditions of life. Transmutations of chromosomes, however, are much more frequently caused by unusual external conditions, this being especially the case in the formation of polyploids.

Such random mutations in all directions as have been artificially induced by treatment with X-rays exactly fulfil the requirements of Darwin's theory that organisms vary slightly in all directions, a few of the changes which are viable and useful being preserved, while a great number which are detrimental are eliminated by natural selection (fig. 194). So far, these artificial radiation experiments may be

Fig. 194. Different varieties of the moth *Oporabia autumnata*, showing the operation of natural selection. The dark variety is less easily seen on the dark bark of the larch than the light variety (above), while on the light trunks of the birch the light one is less visible (below). Consequently the dark variety is more common in larch and pine woods, the light one in birch woods, since the more visible types are quickly seen and are seized upon and eaten, thus giving rise to a predominance of the type most suited to the surroundings. (After Heslop-Harrison.)

regarded as a complete confirmation of Darwin and his law of Natural Selection, with the further illustration of the possibility that creative evolution has proceeded largely as a resultant of the intermittent action of short wave radiations present in the terrestrial atmosphere, some of which may be of cosmic origin.

In this way it is conceivable that a steady evolutionary process has been maintained through the mechanism of the chromosomes and genes under the strict control of natural selection. For countless ages this evolution has steadily proceeded, new mutations and transmutations constantly arising and new species evolving from and in time replacing the old, the genes and chromosomes providing the living mechanism of creative evolution.

Experiments show that the most useful unit of creative evolution is the genetical species, which is usually larger than a Linnean species and often equal to a generic section or sub-genus. Given evolution by descent with modification, now accepted by all expert biologists, creative evolution, as defined above, consists of changes from old unit-species to new unit-species. We have seen that these changes involve corresponding changes in the chromosomes and genes of the new species, so that the living mechanism of the chromosomes and genes through which this change of unit is brought about is in reality the mechanism of creative evolution. In regarding the genetical species as the unit of creative evolution it will be understood that all the higher groups of species such as genera, families, orders, classes and phyla are represented, since all these were originally species in secular time just as some of the species of to-day are the incipient genera and families of the future. So that the genetical species is a true unit of evolution, and one of the most important results of recent researches in Genetics is the placing of the species concept on a basis of experimental reality.

From this aspect we may see the whole possible range of evolution, though since the ancient species have for the most part vanished, we can only see the course evolution followed by a study of the fossil species in the rocks, which, except in a few exceptional circumstances, are of necessity fragmentary in the extreme. Until the forms of life became of sufficient complexity to form some sort of solid structure there was no possibility of their delicate bodies being preserved in the sediments laid down. The slight structure of pre-cellular organisms would be crushed out of existence, and it is only in a few very exceptionally formed layers of rocks that some of the lower species of bacteria and algae have been preserved and one realises that long ages had gone before, while the ultra-microscopic progenes were becoming more complex to form the sub-cellular and the more perfected one-celled forms, some of which in their turn carried on to form the more differentiated but still extremely primitive creatures with few cells. Many more millions of years were to elapse before the first creatures with signs of the formation of some sort of skeleton appeared, and many more before this skeleton became of sufficient strength to withstand the ever-increasing pressure of new layers of sediment and to form good fossils (see p. 323). It is only in recent years that the full force of this has been realised and that we have learned how very long life must have been present on our globe before any fossils were laid down.

Fortunately the records of the rocks, fragmentary though they are, have given ample and convincing evidence that evolution on the whole has proceeded through ages of time from organisms more or less simple in structure to organisms of great complexity.

The immense amount of work accomplished by palaeontologists in their studies of fossils leaves no doubt as to the workings of creative evolution in the past. Many records are extremely fragmentary, but in some cases we have an almost perfect record of the evolutionary changes which have taken place. For instance, on the North American continent there has been found a remarkable series of fossils extending down into the rocks, each layer of rock standing for a long period of time, and we find a remarkable story of the evolution of the horse. In its early stages it is found as a tiny beast with four toes (*Eohippus*) (fig. 195), then it develops in size with only three toes (*Mesohippus*); later it appears larger still with the two side toes reduced, until finally the modern horse (*Equus*) appears with only one toe, but the remains of the other two are still visible when we examine the bones of the foot in a skeleton or the splint bones often thrown up by a young horse. The elephant also shows a similar series from a small animal with a long snout or nose (*Moeri-*

therium) through successive changes of larger and longer-nosed species until we get the huge animal (*Elephas*) of the present day with its distinctive long and sensitive trunk (fig. 197).

In these two parallel instances of the evolution of the horse and the elephant we have interesting examples of evolution by gains and losses of characters. In both cases there is an increase of size and complexity,

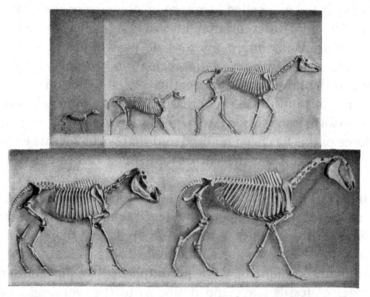

Fig. 195. Photographs of fossils found in successive layers of rock, showing the evolution of the horse. Upper left, the four-toed *Eohippus*, next the three-toed *Mesohippus*, on the right *Meryc-hippus* with the side toes reduced. Below on the left the one-toed *Equus Scotti* and on the right the modern horse. This course of evolution has taken some fifty million years from the *Eohippus* in the early Eocene Period to the present day. (After Loomis.)

e.g. in teeth, trunk and tusks, which in the case of the elephant has resulted in the gain of the trunk and tusk characters while in the horse evolution has resulted in the loss of three of the primitive toes. That all these changes took place by changes of the genes and chromosomes from time to time, there can be no doubt after we have seen the effect of such changes in the genetical experiments carried on to-day.

The scope of this book is not such that we can follow up in detail the great paths of the evolution of Vertebrates, from fishes to amphibians and reptiles, from reptiles to birds and mammals and through the mammals to the conceptual mind of man, of which we have ample evidence in the rocks (pp. 323 and 324). Suffice it to say that evolution has

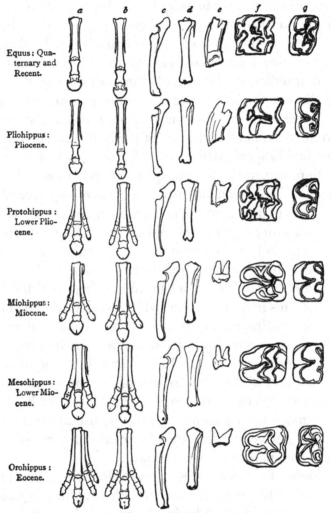

Fig. 196. Evolution of the feet and teeth of the horse. *a*, bones of fore-foot; *b*, bones of hind foot; *c*, radius and ulna; *d*, fibula and tibia; *e*, roots of a tooth; *f*, *g*, crowns of upper and lower teeth. (After Marsh.)

progressed like the growth of a tree, or, perhaps more properly speaking, like a clump of trees, for there are many main stems which have shot out into side branches in all directions from which have come ever new branches and twigs. Many species died out by the way, such as the huge reptiles that existed millions of years ago which became extinct from the very size of their bodies and the weight of their armour, to the growth of which had been sacrificed the size and complexity of their brains. Not only did they overtop the limits of perfect equilibrium by their immense size, but the smallness and simplicity of their head brains, and consequent lack of intelligence, later made them an easy prey to their more lithe and intelligent rivals, the ancestors of our present-day mammals. Many other forms died out by parts of their bodies becoming over-developed to the detriment of the rest of the body, and it is evident that it is not to the highly specialised species that we may look for a continuation of evolution, but to the less conspicuous, generalised species.

Those forms of life which have become extremely specialised for a particular type of environment are not plastic enough as a rule to change with sufficient rapidity to meet the requirements of a rapidly changing environment, especially in the case of plants, which have not the mobility of animals which enables the latter to choose their own environment to a certain extent. The more humble generalised forms are able to radiate new forms in all directions which will be capable of surviving under the new conditions and provide a new phase of evolutionary development which, becoming over-specialised in its turn, will in all probability give place to another. In this way throughout the ages first one and then another of the great groups of protists, plants and animals have ruled and dominated the world, each in its turn being replaced by newer and more plastic forms, advancing usually from some comparatively lowly but generalised member of the progressive group and culminating in the mammal primate man.

Further evidence for the progress of evolution lies in the study of embryology, that is, the study of the growth and development of the fertilised egg-cell into an adult body. In mammals we find represented in the embryonic stages many of the phases through which their ancestors passed in climbing the tree of evolution. To begin with, all start life as a single cell like the uni-cellular forms. This divides and forms a mass of cells (usually a hollow sphere in which forms a gastric cavity) resembling the simpler forms of multi-cellular animals. In later stages the embryo resembles that of a fish, the cavity between the mouth and nose being pierced by a series of gill slits. In the human embryo

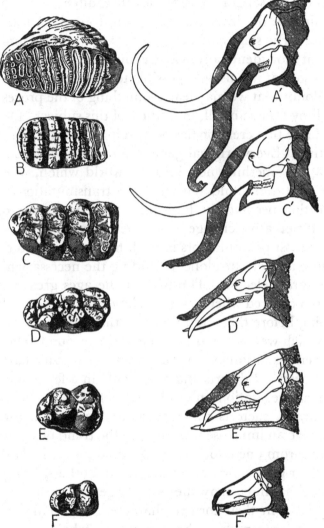

Fig. 197. Evolution of the elephant (teeth and tusks). F, *Moeritherium*; E, *Palaeomastodon*; D, *Trilophodon*; C, *Mastodon*; A, *Elephas*. This evolution is coterminous with that of the horse (figs. 195, 196), *Moeritherium* existing in the Eocene Period. (After Lull.)

later stages resemble that of an ordinary four-footed animal, but the fingers and toes are webbed like an amphibian frog and at this time there is a well-developed tail. Later still there is a complete covering of hair like the hairy skin of an ape, and we all know how the new-born baby has its big toe separated from the others after the manner of a hand and as we find in the hind feet of an anthropoid. All animals follow through the early stages of development in a very similar manner (fig. 198) but not identically, each genus and family showing minor peculiarities so that it is usually possible to distinguish the embryos, and there is no doubt that we may learn something of the processes of evolution, though by no means all, in a study of this subject. On every hand we find evidence that creation has been going on and is going on through all secular time. Every original gene mutation, however small, involves the bringing of something new into the world which, when combined with other mutations and chromosome transmutations, may under natural selection become the forerunner and progenitor of a definite and important creative change, e.g. a new species.

One of the most potent factors in evolution has undoubtedly been the ever-changing face of our globe, providing the necessary milieu for the action of natural selection. Throughout the ages great periods of depression have occurred when great seas have inundated the land, sweeping everything before them and isolating the various species of animals and plants which were saved on the high-lying ground. Then there have been other periods of intense dryness when the greater part of the continents have been hot deserts and again only in a few more fertile parts could a normal vegetation and animal life flourish. Any form of geographical barrier, such as seas, deserts and mountain ranges, arising will always have an immense influence on the trend of evolution, species being isolated from one another and gradually or rapidly diverging by changes in their genes and chromosomes, and others, being driven into unaccustomed habitats, may die out altogether or survive only by a complete change in their former characters, especially in the case of plants which are not mobile. Sometimes the whole globe has enjoyed a pleasant equable climate extending even to the poles, and these periods may be perhaps regarded as the normal condition. During these "golden ages" vegetation and animals have spread to every corner of the earth and grown and multiplied exceedingly. These periods of calm have been interrupted by great volcanic upheavals which have filled the air with gases and thrown up new mountain ranges and islands, causing a redistribution of oceans, rainfalls, and so forth, which have given rise to entirely new habitats and climates. More disturbing still, perhaps, have been the great Ice Ages which have occurred periodically, in which the poles or other areas have become covered with ice, which has gradually or in some cases rapidly spread southwards (or north-

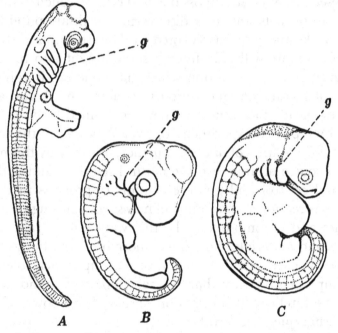

Fig. 198. Corresponding stages of development in the embryos of A, Shark, B, Fowl, and C, Man. g = gill slits. (After Scott.)

wards), swamping everything before it and casting a cold breath over all the land under its influence. During the last Ice Age, from which we are only now emerging, the greater part of Europe and North America was engulfed by the ice, which had a great effect on the destruction or redistribution of plants and animals, causing the relatively rapid creation of many new species, and it is considered to have been the main factor in the evolution of the present species of man (see p. 330).

Given all these changes of conditions of life and the consequent effect of natural selection on the multitudinous mutations and transmutations, arising maybe chiefly from cosmic radiations, it is not surprising that the animal, plant and other living kingdoms have passed through great changes of genic constitution with the consequent changes of characters.

In this book it has not been possible to do more than touch the fringes of this vast subject, which is daily becoming more and more complicated and growing to such an extent that those of us who devote our whole time to research, find it impossible to keep in touch with the vast accumulation of facts and new discoveries which are published in the numerous books and journals devoted to these researches. It is hoped, however, that many will gain from it some insight into the wonderful mechanism of creative evolution which has gone on, is going on, and will no doubt go on from generation to generation. A vast amount of research lies ahead of us, but every new fact gained by observation and confirmed by repeated experiment, opens a new way to the true understanding of the workings of nature. Further, it is only by this means that man can in course of time displace natural selection and with his conceptual mind control and guide evolutionary progress in the way he desires, by the creation and selection of new species through the changing of the genes and chromosomes. In this way he can gain for himself better foods and materials of all kinds, more beautiful things to admire, and eventually, if he so wills, create a race of super-men to enjoy life in a more comprehensive way than we can hope to do, and at the same time gain the intellectual satisfaction derived from power over nature and the eternal quest for Truth.

CHAPTER XIX

EVOLUTION OF MAN AND MIND

Ancestral and Evolved Groups	Geological Periods	Estimated Age in Years	Evolution of Mind
Matter on planet Earth (protons, electrons, atoms, molecules)	Azoic	2000 million	Electrical action
Protogenic molecules	Eozoic	1000 million	Autocatalytic reactions
Progenes and genes	Eozoic	1000 million	Autogenic responses
Schizophyta (bacteria and blue-green algae)	Proterozoic	900 million	Pre-cellular responses
Protozoa	Protozoic	800 million	Uni-cellular tropic responses
Invertebrate animals	Late pre-Cambrian	600 million	Multi-cellular tropic responses
Vertebrate animals	Ordovician	500 million	Vertebrate reflexes
Fishes	Silurian	450 million	Piscal reflexes
Amphibians	Devonian	400 million	Amphibian reflexes
Reptiles	Carboniferous	300 million	Reptilian reflexes
Mammals	Early Triassic	200 million	Mammalian reflexes
Placentals	Early Cretaceous	100 million	Placentalian reflexes
Insectivores	Middle Cretaceous	80 million	Insectivoral reflexes
Primates	Late Cretaceous	60 million	Primatal reflexes
Lemuroids and tarsioids	Early Eocene	50 million	Lemurian and tarsian reflexes
Catarrhines	Eocene	40 million	Catarrhinal reflexes
Pre-anthropoids (*Propliopithecus*)	Early Oligocene	30 million	Pre-anthropoid reflexes
Rusinga anthropoid	Miocene	20 million	Anthropoid reflexes
Pre-human anthropoids (*Dryopithecus*)	Late Miocene	12 million	Pre-human reflexes
Taungs ape-man (*Australopithecus*)	Early Pliocene (?)	10 million (?)	Human teeth and jaw
Eolithic men (Cantalian)	Early Pliocene	10 million	*Primitive human conceptual mind*
Eolithic men (Kentish)	Middle Pliocene	6 million	Kentish eoliths
Eolithic men (Rostro-carinates)	Late Pliocene	2 million	Rostro-carinate eoliths
Piltdown ape-man (*Eoanthropus*)	Late Pliocene	1 million	Piltdown flint and bone implements
Foxhall men (Ipswich)	Late Pliocene	1 million	Foxhall industry (fire and cookery, with bone implements)
Peking ape-man (*Sinoanthropus*)	Early Pleistocene	750 thousand	Traces of speech (fire and bone and stone implements)
Oldoway man (*Homo sapiens*) and Kanam jaw	Early Pleistocene	750 thousand	Chellean and pre-Chellean implements

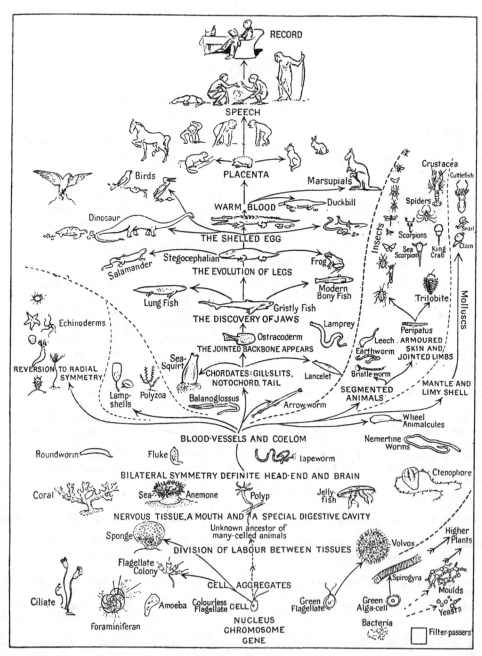

Fig. 199. Creative Evolution.

(From *The Science of Life*, by H. G. Wells, Julian Huxley and G. P. Wells.)

Ancestral and Evolved Groups	Geological Periods	Estimated Age in Years	Evolution of Mind
Heidelberg man (*Palaeanthropus heidelbergensis*)	Early Pleistocene	750 thousand	Slow speech
Kenya men	Gunz Pleistocene	600 thousand	Acheulian implements
Java ape-man (*Pithecanthropus*)	Middle Pleistocene	500 thousand	Almost erect
Java Solo man (*Homo soloensis*)	Middle Pleistocene	500 thousand	Primitive Neanderthal and Rhodesian
Cromer men	Gunz-Mindel Pleistocene	500 thousand	Cromerian implements
Ehringsdorf man (*Homo neanderthalensis*)	Mindel Pleistocene	400 thousand	Pre-Mousterian implements
European Chellean men	Mindel Pleistocene	400 thousand	Chellean industry
European Acheulian men	Mindel-Riss Pleistocene	200 thousand	Acheulian industry
Neanderthal men (*Homo neanderthalensis*)	Riss Pleistocene	150 thousand	Mousterian industry
Rhodesian man (*Homo rhodesiensis*)	Late Pleistocene (?)	50 thousand (?)	Neanderthaloid, australoid and negroid
Mousterian man of Mount Carmel (*Palaeanthropus palestinus*)	Wurm I	25 thousand	Mousterian industry
Hunstanton man (*Homo sapiens*)	Laufen retreat Pleistocene	20 thousand	Aurignacian implements

Modern Man (*Homo sapiens*)	Recent Period	B.C.	Evolution of Mind
Aurignacian men	Achen Oscillation	11,500	Aurignacian industry
Cromagnon men	Achen Oscillation	11,000	Early Capsian art
Solutrean men	Buhl I	10,000	Solutrean implements
Magdalenian men	Buhl I	9,500	Magdalenian industry. Late Capsian art
Microlithic men	Buhl III	6,500	Microlithic industry. Final Capsian art
Men of the Orient	Ancylus Lake	6,000	Agriculture and pottery. Stone-grinding and metal work. Spinning and weaving
Mesolithic men of Mt Carmel	Ancylus Lake	5,500	Mesolithic bone craft
Men of the Fertile Crescent	Ancylus Lake	5,000	Badarian culture (Egypt). Susa I settlement (Iraq). Neolithic Cretan culture
Men of Surrupak. Ancient Egyptians	Gschnitz	4,500	"The Flood" in Iraq. Predynastic culture (Egypt)
Ancient Semites, Elamites and Sumerians (Chaldeans)	Littorina Sea	4,000	Civilisations of Kish I and Ur I (Iraq)
Men of Anau	Littorina Sea	4,000	Civilisation of Anau I (Turkestan)

Modern Man (*Homo sapiens*)	Recent Period	B.C.	Evolution of Mind
Dynastic Egyptians, Cretans and Vinca Danubians	Littorina Sea	3,500	Dynasty I (Egypt). Early Minoan civilisation
Babylonians	Littorina Sea	2,000	Kingdom of Babylon
Assyrians	Littorina Sea	2,000	Kingdom of Assyria
Trojans	Daun	1,500	Ancient Troy
Ancient Greeks	Historic Time	1,000	Early poetry in Greece
Ancient Greeks	Historic Time	500	Ionian philosophy
Ancient Greeks	Classical Period	400	Greek philosophy, art, literature and civilisation
Ancient Romans	Classical Period	30	Roman empire
Ancient Romans	Christian Era	A.D.	Roman civilisation. Christianity
Byzantines	Dark Ages	5th century	Byzantine empire
Arabs	Dark Ages	8th century	Arabian science and learning
Europeans	Middle Ages	12th century	Christian universities
Europeans	Middle Ages	13th century	Mediaeval scholasticism
Italians	Renaissance	15th century	Italian renaissance. Pictorial art of old masters. Beginnings of modern science
English	Renaissance	16th century	Elizabethan poetry and drama
Europeans	Development	17th century	Science and mathematics. Telescope and microscope
Europeans	Philosophy	18th century	European philosophy
Europeans and Americans	Materialism	19th century	Experimental and applied science. Pure and applied mathematics. Philosophy of evolution
Civilised World	Scientific Research	20th century	Age of scientific research. Pure mathematics. Principles of relativity and indeterminacy. Experimental evolution

CHAPTER XX

SPECULATIONS

WE have seen how creative evolution has proceeded in the past by the gradual growth of complexity in organisms as a direct consequence of the increase in complexity of their gene content brought about by the mechanism of mutations and transmutations of the genes and chromosomes, so that new species have evolved, which in the course of ages of time, under the control of natural selection, have become genera, families, orders, classes, phyla, kingdoms and super-kingdoms.

It is estimated that the super-kingdom of Organic Life first appeared approximately halfway between the origin of the earth as a planet from the sun and the present time. The beginnings of organic life are lost in the obscurity of the past and in the ultra-microscopic nature of the basis of life. Since all macroscopic and microscopic life known to us experimentally is found to contain or carry genes in one form or another, and experiments with higher organisms have demonstrated that the gene is the unit and basis of living characters, it is reasonable to infer that life first arose in the form of primitive genes or protogenes, most probably of the nature of autocatalytic molecules arising from those half-alive self-reproducing chemical ferments which seem to follow the formation of certain configurations of carbon and other atoms of matter. From these humble beginnings the gene has built up the edifice of life as we know it, leading to its highest and most complex form, Man.

In relatively recent times the super-kingdom of Conceptual Mind arose in the family of mankind associated with a complex brain-structure and nervous system which had developed through the ages as a gradual evolution of life and matter. From the electrical actions of protons, electrons and atoms and the autocatalytic reactions of genic molecules we find a regular sequence of the evolution of mind through the tropic responses of pre-cellular, uni-cellular and multi-cellular organisms to the reflex mind of vertebrate animals, mammals, primates and anthropoids, until finally conceptual mind arose in certain genera of Pliocene ape-men and developed later in our own genus *Homo*, with the most rapid progress in our own species *H. sapiens* during the last eight thousand years (see p. 325).

In the natural course of creative evolution it is reasonable to conclude that these creative processes will continue and that new species of the kingdoms of animals, plants and protists will arise on the earth in the future as in the past. Looking into the distant future, we may infer that new kingdoms of living organisms other than animals, plants and protists will arise and that in the course of time a succession of new super-kingdoms will gradually come to pass, each great step surpassing and transcending conceptual mind to the same degree and extent that mind now transcends life and life surpasses matter.

It is interesting to speculate on the nature of these great creative steps and to consider the possibilities as to when they may arise. With regard to the question of the nature of the next great creative step transcending mind, recent experimental research provides us with the definite answer that we cannot expect to know beforehand precisely what it will be. We have seen that one of the distinguishing features of creative evolution which differentiate it from other kinds of evolution is the novel, indeterminate and unpredictable nature of its genetical progress. That this must inevitably be so is evident in the findings of this book that the course of creative evolution in living nature has been shaped and guided in the higher organisms by at least four different vital processes, all of which in their action are random variables, namely: Mutation, Transmutation, Sex and Natural Selection. The random mutations of genes and the chance transmutations of chromosomes appear, on experimental evidence, to be caused by atomic or other disturbances due to short-wave radiations and other causes, producing at random every possible kind of hereditary variation. The function of sex, in the higher organisms, serves to combine and recombine at random and to fix these mutations and transmutations in different individual organisms, while the constant action of natural selection determines their survival and consequently the progressive adaptation of the mutants and transmutants to the changing conditions of life. Natural selection, being contingent, is locally random in its action according to the particular conditions of environment which happen to be present during the fertile life of the surviving organism, whether it be a gene, a protist, a plant or an animal. Of the four prime factors concerned with the processes of creative evolution, natural selection has been the final arbiter, and though locally random and contingent in its action it has inevitably made for general progress in creative evolution.

It is evident that when four or more random variables are concerned in the creation of new species in nature, it is impossible to foresee or to foretell the precise nature of the future species, and if this is the case with small and relatively insignificant specific changes, the impossibility of foreseeing or predicting the precise nature of such large super-kingdom changes as life from matter, or mind from life and matter, is manifest.

The precise nature of the next great step in creative evolution is therefore unpredictable. All one can say is that in all probability in the future, as in the past, the new step when it comes will be genetically based on the old steps, mind, life and matter, and hence may with high probability be expected to appear in the descendants of Modern Man, *Homo sapiens* (who is the sole surviving species of the family Hominidae in which conceptual mind first appeared), rather than in the descendants of any other species. In view of the progressive nature of creative evolution and of man in the past we may reasonably expect it to be progressive in the future and consequently may infer that the next great step will be more complex, more versatile and of a higher intrinsic value than our present mind, life and matter.

The question as to when the next great step in creative evolution will arise precipitates us at once into deep waters. The processes of small steps in the evolution of species in nature are usually extremely slow, and such large steps as the evolution of the super-kingdoms of life and mind measured in years of time are almost beyond human comprehension. One of the most remarkable features of modern science is the rapid expansion in recent years of the scientific estimates of the age of man, life, the earth and the universe. Nor are these estimates by any means wild guesses, since all are definitely based on the experimental data of radiation and on convergent astronomical, physical, geological, palaeontological and archaeological evidence. The experimental demonstration of the ultra-microscopic living gene as the unit and basis of life inevitably expands the biological estimate of the age of life on the earth. The earliest abundant fossil remains of protists, plants and animals are found in the Cambrian rocks of the Paleozoic Era which are radium-dated in round numbers as about 500 million years old. The organic remains found there consist of such relatively complex plants and in-vertebrate animals as Algae, Worms and Trilobites. In the older rocks of the Pre-Cambrian Period, fossil remains become rarer, more scattered and are not so well preserved. They are found through the Torridonian Period and into the upper layers of the Charnian rocks of the Proterozoic Era, dating back in round numbers to about 600 million years when they cease to appear. These fossils must have had a considerable evolu-tionary period behind them, apart altogether from the hosts of minute microscopic uni-cellular Protozoa, Algae and the pre-cellular Algae and Bacteria which for the most part would be too small, soft and fleeting to be recorded in the rocks. Behind these would be the untold millions of ultra-microscopic organisms from the larger filter-passers to the protogenes on the borderland of life, whose presence in the rocks, even if preserved, would remain undetected.

In the present state of knowledge, therefore, we may safely estimate that the age of life on the earth can hardly be less than 1000 million years. With regard to the age of mankind and the first manifestations of the human conceptual mind, our own species, known as Modern Man (*Homo sapiens*), so far as our present knowledge goes, is about 750 thousand years old, dating from the oldest deposits of the Pleistocene Period. The earliest skeleton found is that of Oldoway Man, first discovered in 1913 by Dr Reck of Berlin in East Africa at Oldoway, Tanganyika, and confirmed by the discovery of the Kanam Jaw by Dr Leakey in November, 1931. There has been some doubt about the age of this speci-men, but the various animal remains found in the same layer corre-spond with the period, and the human implements found with them have been definitely identified as Chellean, while in the layers below remains of the elephant *Deinotherium* (usually Miocene) were found together with human implements of the Pre-Chellean type. In the layers above were found implements of the later Acheulian type together with remains of *Elephas antiquus* and other animals dating from the 1st pluvial period of the Pleistocene up to the lower middle Pleistocene. Another species of the genus *Homo*, Neanderthal Man (*H. neanderthalensis*), lived in the Pleistocene Period, the earliest remains of which are probably those of Ehringsdorf Man, which are estimated to date from the Great Mindel glaciation of about 400 thousand years ago. Other genera of the family of Mankind (Hominidae) lived in the Pleistocene and Late Pliocene (see p. 323), the earliest of which was most probably the Piltdown Ape-Man of Sussex (*Eanthropus dawsonii*). The remains of his skull, found with implements of flint and bone, are now estimated to be Late Pliocene and about a million years old, although previously considered to belong to the Pleistocene Period. Long before this, however, right through the Pliocene Period, are found more or less rude flint and stone implements known as eoliths which most authorities now conclude were fashioned by human hands. The oldest of these, known as the Cantal eoliths, certainly suggest an artistic purposefulness which involved considerable manual and digital dexterity and indicate the possession of a mind which had arrived at the stage of conceptual consciousness. So far as we know, no other animal, mammal, primate or anthropoid ape has or had the ability to conceive, manufacture and use such tools and weapons, and we are forced to the conclusion that these eolithic implements represent the first expressions of the conceptual mind in mankind. Eolithic man was probably an ape-man belonging to a genus of the Hominidae more primitive than *Homo* and more advanced than any of the anthropoid apes of the family Simiidae. The age of the Cantal eoliths is generally accepted as Early Pliocene or Late Miocene, which according to the latest calculations represents an age of about 10 million years, and for the present we may accept this date as a round figure estimate of the age of conceptual mind in man. This was the period in creative evolution when man as a human being became segregated from the pre-human stock of anthropoids known to palaeontologists as *Dryopithecus*, a large and variable genus (or perhaps more correctly a family Dryopithecae) distributed over the greater part of the Old World during this period. Most authorities agree that this generalised stock of pre-human anthropoids gave rise to the two distinct families of Hominidae (Mankind) and Simiidae (Anthropoid Apes), each of which diverged from one another during the last 10 million years with the results that can be seen to-day. On the one hand the generalised *Dryopithecus* progressively developed into the widely distributed *Homo sapiens* (Modern Man) of the Hominidae with all the generalised potentialities of a conceptual mind, while on the other hand *Dryopithecus* developed in other directions into the chimpanzee, gorilla, orang-outang and the diminutive gibbon of the Simiidae whose specialised characters and reflex mind have led them to the blind alley and dead end of a circumscribed habitat, and their ultimate survival will depend largely on the will and conceptual mind of the superior species, Man.

In accordance with these estimates we may consider the age of conceptual mind in mankind to be about 10 million years and the age of life on the earth to be about 1000 million years.

Jeans estimates the age of the earth as a planet to be in round numbers 2000 million years. On this basis it would appear that matter had a solitary reign on earth of 1000 million years before life appeared and that matter and life had a joint existence of nearly 1000 million years before mind appeared in mankind in the form of conceptual consciousness.

In the genetical evolution of mind through life and matter there are definite indications that the most recent and highest term of the three —mind—is gradually increasing in influence as creative evolution proceeds, while the oldest and lowest term of the three—matter—is gradually decreasing in influence. Even now, after the relatively short time of about 10 million years since conceptual mind appeared, it is evident that by the exercise of the human mind, human selection is gradually displacing natural selection in many respects and there is no doubt that through the applications of experimental science, man, as time goes on, will be able more and more to control and change his environment and his conditions of life as well as those of other organisms. Man's increasing control over matter is due almost entirely to the exercise of his conceptual mind in scientific experiments.

We see the results of the increasing influence of mind over matter in the remarkable mechanical achievements of the present century. If experimental genetics progresses in the future as rapidly as it has done in the last decade, it will not be long before man will be able to induce and control the creative evolution of other species and, if he so wills, to induce, direct and guide the creative evolution of his own species.

There are therefore definite indications in the present century that under the influence of the conceptual mind of man, creative evolution may be speeded up, and it is possible that the next great step surpassing mind may be evolved in much less time than the 1000 million years estimated for the evolution of life and mind. It may be that in the next 1000 million years several great steps successively surpassing conceptual mind and one another will be creatively evolved, and that each step will speed up the evolution of its successor.

A new species in nature is always based genetically on an old species, and in the course of evolution the influence of the ancestral species becomes less and less until in course of time the old specific characters are obliterated in the modern species, which bears a new set of specific characters. It is true that the old generic and family characters remain in the new species, but in the course of evolution even these disappear in new genera, families, orders, classes, phyla and kingdoms. The super-kingdoms of matter, life and mind may be regarded as species writ large in time and value in the course of creative evolution. We have seen that mind is gradually increasing in influence at the expense of matter, and it is reasonable to infer that in course of time when the next great step in creative evolution appears, the influence of matter will have been considerably reduced and displaced by mind and its unknown successor. The further inference is that in the course of long ages and perhaps several great creative steps, successors of man will be evolved in whom the influence of matter has been almost, if not entirely, obliterated, and a less material type of being will arise, utterly different from the present human species, scarcely human save in mind and thought but on a higher intellectual plane. Such an independence of matter would enable the more adventurous of our far away descendants to leave the earth and to visit and people other planets in our solar system or other stellar systems of our universe and even other universes if they exist.

Many modern spiritualists and some men of science believe in such an ethereal human existence supervening immediately after death, and although the experimental evidence presented in favour of this belief is obscure and difficult to understand, it is not impossible that with the evolution of conceptual mind in mankind there came also the potentiality of its survival after death. There seems to be no valid scientific objection to the belief in a future existence in the form of pure thought or spirit, and such beings may exist in other parts of the universe as a product of creative evolution in other planets.

In view of such speculations it is interesting to compare the latest astronomical picture of the future of the universe portraying the ultimate and final dissipation of matter into radiation, with the biological picture of the evolution of matter into life, mind and many higher values in the distant future.

On the one hand, we have an astronomical picture of our universe as a universe finite in time and space which, according to the laws of thermodynamics, by the rapid "increase of entropy", is fast proceeding to a final dissolution and death. This is a sad and gloomy picture which is only slightly relieved by the consolation that notwithstanding its ever-increasing rapidity of approach, the end is not expected to come for some millions of years.

On the other hand, we have a biological picture of the creative evolution of matter into life, mind and many higher values during this future period of millions of years with the possibility of an infinite intellectual or spiritual existence beyond. This is a much brighter and more cheerful picture of the future and the question arises whether these two apparently antithetical pictures can be reconciled and harmonised. This is possible once it is realised that the astronomical picture is concerned only with the future of matter while the biological picture represents the future of life and mind. As the latest scientific hypothesis of the future of matter and the universe, based on the laws of thermodynamics, the astronomical picture must naturally be

accepted for the time being, but since life and mind represent a later
stage of evolution than matter the biological picture should be super-
imposed on the astronomical picture. In our biological picture we have
seen that there are definite indications of the decreasing influence of
matter on life and mind, and the natural inference is that in the course
of time the influence of matter on life and mind will gradually become
obliterated and replaced by the influence of mind and its transcendent
successors in the great steps of creative evolution. Long before the time
comes for the annihilation of matter and the universe, the more primitive
matter will have lost its influence on life and mind and their successors,
and the remote successors of man may be free and independent of
matter with infinite possibilities of future progress notwithstanding the
complete dissolution of the universe.

The superimposition of the biological picture on the astronomical
picture thus presents a finite universe of matter in finite space-time
creatively evolving into infinite progressive values far surpassing and
transcending our present values of matter, life and mind.

At the present time, owing to the advent of conceptual mind, our
main interest and concern lie in the psychological evolution of mankind.
We have seen that conceptual mind in mankind is about 10 million
years old and that since it first appeared in the form of the human industries of tool and weapon making there has been a gradual development through the ages. Up to relatively recent times this development
was exceedingly slow, but after the appearance of the new species *Homo
sapiens*, known as Modern Man, the progress of conceptual mind became
relatively rapid. Human industries of wood, stone, flint and bone became more complex and artistic and the Capsian art of mural painting
and modelling in clay was soon highly developed. Up to about ten
thousand years ago, man was a hunting animal like his predecessors the
Cave Men, but about this time, or soon after, he began to settle down to
domestic industries and agriculture, to the cultivation of food plants and
the domestication of animals. Recent discoveries indicate that about
eight thousand years ago the era of modern civilisation began to dawn,
probably in the Fertile Crescent between North Syria and the Upper
Valley of the Euphrates. The earliest known civilised settlements are

those of Antediluvian Susa in Iraq, the Badarian Culture in Egypt on the banks of the Nile and the Neolithic Culture of Crete and the Aegean Islands, all of which are estimated to be about seven thousand years old. These cultures include fine pottery, copper tools and ornaments, finely flaked implements of Solutrean design, ground stone axes, ivory figures, woven linen, cultivated grain and the domesticated dog.

After the "Flood" in Iraq there followed the civilisations of Kish and Ur about six thousand years ago, side by side with that of Anau in Turkestan. About five thousand years ago the First Dynasty of Egypt and the Early Minoan Period of Crete arose side by side with ancient Babylon, Assyria and the Danubian Settlers at Vinca.

These ancient civilisations waxed and waned in their turn, ultimately culminating in the high standard of conceptual mind reached in Ancient Greece from three thousand to two thousand years ago, from the time of Homer and the Ionian philosophers to that of Plato and Aristotle and the extension of Greek culture to Alexandria, the Orient and India. Immediately after came the advanced practical civilisation of the Roman Empire, which in a few centuries declined and led to the Dark Ages of the West and the lesser Byzantine Empire of the East.

The advent of three great religions in the East between the seventh century B.C. and the seventh century A.D., Buddhism in India, Christianity in Palestine and Mohammedanism in Arabia, profoundly influenced the known world, giving rise to new ethical standards and leading to the ecclesiastical culture of the Middle Ages with its distinctive religious poetry, literature and art and its restricted philosophy and science. The Arabian conquests in the East in the seventh century A.D. led to the establishment of Arabian science and philosophy from India to Spain and was maintained until the end of the eleventh century.

In the twelfth century, Christian Universities arose in Europe and with them came a revival of Greek thought through translations of the Arabic into Latin, and mediaeval scholasticism reached its highest level in the thirteenth century. The fifteenth century brought the Italian Renaissance with its remarkable intellectual and artistic developments, comparable only with those of Ancient Greece and those of our own day. At this period pictorial art in the work of the old masters reached a standard that has not been surpassed before or since. The discovery of America and the development of the art of printing widened the mental outlook and loosened the shackles of mediaeval and ecclesiastical scholasticism.

Thought once more became free, as in ancient Greece and in our own day, leading to wider developments of philosophy and science and a return to nature. In these favourable circumstances modern experimental science began to develop, leading up through three centuries to the remarkable scientific achievements of the nineteenth century and the notable mechanical and engineering developments of the present century.

This development and extension of conceptual mind in the nineteenth century has also given rise in the present century to a radically new natural philosophy based on experimental science and mathematics. The development of pure mathematics has enabled Einstein and others to propound revolutionary views of the relativity of the universe surpassing those of Newton and the earlier mathematicians, and has enabled the physicists to present and establish new concepts on the nature and constitution of matter. The discovery of the electron in 1899 by Thomson showed the complex nature of the atom and completely changed our conceptions of the fundamental basis of matter. The discovery of the proton by Rutherford demonstrated further complexities in the atom and has done much to solve the problems of the constitution of matter. Incidentally, by the transmutation of chemical elements through radiation, he has presented the old alchemist's problem in a new light. The work of Eddington, Jeans and others in astrophysics has considerably extended and broadened our conceptions of the universe. The peculiar experimental behaviour of the electron has led to the new and important Quantum Theory of Planck, which has revolutionised the science of physics during the last few years, and has provided scientific evidence for Heisenberg's principle of indeterminacy in nature and led to the new mathematics of Dirac.

During the twentieth century the fundamental advances in biology have been even more revolutionary and significant to man than those in physics and have led to the foundation of an exact science of experimental evolution, based on genetics. The publication of the *Origin of Species* in 1859, which heralded the Darwinian Renaissance of the nineteenth century, was immediately followed by the classic experiments of Mendel which, however, only came to light in 1900. Mendel's experimental discovery of the unit of heredity, now known as the gene, has completely revolutionised biology by giving rise to the new and exact science of genetics which in the course of a few years has changed the nature of biology from a largely qualitative observational science to an exact quantitative experimental science.

In the first decade of this century Mendel's discovery of the genes in Peas (*Pisum*) was extended by the experiments of Bateson and others to all kinds of plants and animals, including man, and Mendel's laws of heredity were firmly established. Earlier in the century, De Vries had already established the principle of mutation in the Evening Primrose (*Œnothera*).

In the second decade, Morgan and others in America succeeded in experimentally determining the location and linkage of the genes in the chromosomes of the Fruit Fly (*Drosophila*), which further confirmed the laws of Mendel and provided a physical basis for the genes in the chromosomes of the nuclei of the cells.

In the third decade, four important advances have been made, all of which are based on the genetical laws of Mendel. (1) The discovery of the transmutations of chromosomes and genes leading to the experimental creation of new varieties and species. (2) The experimental creation of gene mutations and chromosome transmutations by the application of X-rays and other short-wave radiations. (3) The discovery that specific and higher group characters are also represented by genes which, after mutation, behave as Mendelian varietal characters. (4) The translation of the Linnean concept of species into terms of genes and chromosomes, thus establishing, in the genetical species, a measurable unit of evolution together with a new exact science of taxonomy founded on an experimental basis.

These recent advances in biology have been dealt with in this book and it is evident that the discovery of the gene as the unit and basis of life, heredity, mutation, transmutation, species and creative evolution has through experimental genetics laid the foundations of an exact science of experimental evolution. These fundamental developments in biology, when they become generally known, will naturally cause considerable changes of view in the kindred branches of physiology, pathology, biochemistry, biophysics, taxonomy and phylogeny and eventually in the more remote realm of philosophy.

The new knowledge has provided the missing link in Darwin's theory of natural selection and thus serves to establish it firmly and finally as a prime law of nature and the ultimate arbiter, though not the primary cause, of creative evolution. Darwin's postulate of the existence of hereditary variations in all directions has not only been experimentally demonstrated, but one of the prime sources of origin of these variations has been traced experimentally to the action of short-wave radiations. Darwin's crude and speculative "gemmules" (pangenes) of the nineteenth century have been experimentally identified as Mendelian genes, and Darwin's hereditary variations in all directions are recognised as the resultants of the mutations of genes and the transmutations of chromosomes.

In the light of recent knowledge the Lamarckian discussions of the nineteenth century on the inheritance of acquired characters have lost their meaning, and if Lamarckism or indeed any other theory of evolution is to be acceptable to biologists of the twentieth century it will be necessary to restate it in terms of genes and chromosomes.

Owing to the indeterminate nature of mutations and transmutations and the local contingency and randomness of natural and human selection, it is impossible to predict the precise nature of future species and consequently to foresee the future of creative evolution. It is at this point that scientific determinism breaks down. The deterministic and mechanistic evolution of the nineteenth century, with its closed circuits and endless cycles, proves to be altogether inadequate to account for creative evolution as we know it in the twentieth century. Closed circuits and endless cycles provide little room for the novelty and progressiveness which are the essence of creative evolution. If we replace the biological unit of creative evolution—the genetical species—by the wider philosophical units of creative evolution—matter, life and mind—the

22-2

inadequacy of determinism becomes even more marked. No deterministic system, scientific or philosophical, could possibly have predicted life from matter or mind from life and matter. It may be, as Eddington suggests, that in this principle of indeterminacy in nature we come up against the reality behind our man-made laws of nature. Our human geometrical minds work naturally in terms of mechanism and the determinate resultants of cause and effect—anything outside that seems unreal to the human conceptual mind. It may be that the principle of indeterminacy, so incomprehensible to us, is a universal principle of thought behind and beyond our human conceptual laws of mechanistic determinism. If so, many mysterious problems find an explanation, if not a solution, in this interpretation.

Jeans pictures our universe as a universe of thought and its creation as an act of thought with space-time as a setting. This seems to involve a prime thinker and creator outside his own creation. In this respect modern scientific speculation returns to the ideas of Plato, who taught the doctrine that in creation, time and space came into being at the same instant, so that if they were ever to dissolve they might be dissolved together, which is an interesting anticipation of Einstein's demonstration of the inseparability of space and time.

The real nature of time is not at all clear, but if we regard creative evolution as creation extending over past, present and future time, then the act of thought, postulated by Jeans as giving rise to creation, might cover the whole of time, past, present and future. Such a conception would go far to explain the progressive nature of creative evolution. If matter consists of electrical events, life of organic events and mind of conceptual events, all originating from and expressions of pure thought, then the genetical continuity of matter, life and mind is more easily understood. Once it is realised that matter has given rise to life and mind, the old idea that matter, life and mind are independent entities passes away and the Cartesian dualisms of matter and life, matter and mind, and life and mind no longer exist. All three are fundamentally and genetically one with different expressions in time, forming a monistic trinity with a common basis and origin in pure thought, which after all may be only another name for spirit.

The great principle of indeterminacy in nature may thus be tentatively interpreted as the natural mode of action and manifestation of pure thought or spirit in matter-life-mind, while determinism, which is none the less real, represents an inner and secondary principle based on the geometrical mode of action of the present human conceptual mind.

It may be that the discovery in our time of scientific indeterminacy and relativity foreshadows the coming of the next great step in the creative evolution of mind, far surpassing and transcending the present conceptual and deterministic mind of man.

REFERENCES

(1) BOOKS

BABCOCK, E. B. and CLAUSEN, R. E. (1918). *Genetics.* New York.

BATESON, W. (1913). *Mendel's Principles of Heredity.* Cambridge.

—— (1913). *Problems of Genetics.* Yale.

BERGSON, H. (1923). *L'Évolution Créatrice.* (26th ed.) Paris.

CASTLE, W. E. (1932). *Genetics and Eugenics.* Chicago.

CREW, F. A. E. (1925). *Animal Genetics.* London and Edinburgh.

—— (1927). *Genetics of Sexuality.* Cambridge.

—— (1927). *Organic Inheritance in Man.* London and Edinburgh.

DARWIN, C. (1859). *Origin of Species.*

—— (1868). *Animals and Plants under Domestication.*

DAVENPORT, C. B. (1912). *Heredity and Eugenics.* London.

DE VRIES, H. (1901). *Die Mutationstheorie.* Leipzig.

D'HERELLE, F. (1930). *The Bacteriophage.* London.

EAST, E. M. and JONES, D. F. (1919). *Inbreeding and Outbreeding.* Philadelphia.

EDDINGTON, Sir ARTHUR (1928). *Nature of the Physical World.* Cambridge.

EIMER, G. H. (1890). *Organic Evolution.* (Eng. trans. by J. T. Cunningham.) London.

ELLIOT SMITH, G. (1931). *Man's Ancestors.* London.

FISHER, R. A. (1930). *Genetical Theory of Natural Selection.* Oxford.

FORD, E. B. (1931). *Mendelism and Evolution.* London.

GALTON, Sir FRANCIS (1869). *Hereditary Genius.* (2nd ed. 1892.) London.

GARDNER, A. D. (1931). *Microbes and Ultramicrobes.* London.

GATES, R. R. (1915). *Mutation Factor in Evolution.* London.

—— (1929). *Heredity in Man.* London.

GUYER, M. F. (1928). *Being Well-Born.* (2nd ed.) London.

HAECKER, V. and ZIEHEN (1923). *Vererbung und Entwicklung Musikalischen.* Leipzig.

HURST, C. C. (1925). *Experiments in Genetics.* Cambridge..

JEANS, Sir JAMES (1931). *The Mysterious Universe.* Cambridge.

JUDD, J. W. (1911). *The Coming of Evolution.* Cambridge.

KEITH, Sir ARTHUR (1927). *Man's Origin.* London.

LAMARCK, J. B., Chev. de (1809). *Philosophie Zoologique.*

LOTSY, J. P. (1916). *Evolution by Hybridisation.* The Hague.

MATSUURA, H. (1929). *Bibliographical Monograph on Plant Genetics.* Tokyo.

MORGAN, C. L. (1927). *Emergent Evolution.* London.

MORGAN, T. H. (1914). *Heredity and Sex.* (2nd ed.) New York.

—— (1919). *The Physical Basis of Heredity.* Philadelphia.

—— (1925). *Evolution and Genetics.* Princeton, U.S.A.

—— (1926). *Theory of the Gene.* Yale.

MORGAN, T. H., STURTEVANT, A. H., MÜLLER, H. J. and BRIDGES, C. B. (1915). *Mechanism of Mendelian Heredity.* New York.

NEWMAN, H. H. (1921). *Evolution, Genetics and Eugenics.* Chicago.

OSBORN, H. F. (1908). *From the Greeks to Darwin.* London.

—— and Others (1928). *Creation by Evolution.* New York.

PEAKE, H. and FLEURE, H. J. (1927). *The Corridors of Time.* I, II and III. Oxford.
PUNNETT, R. C. (1927). *Mendelism.* (7th ed.) London.
SCHÜRHOFF, P. N. (1926). *Die Zytologie der Blütenpflanzen.* Stuttgart.
SHARP, L. W. (1926). *Cytology.* New York.
SMUTS, J. C. (1926). *Holism and Evolution.* London.
WELLS, H. G., HUXLEY, J. and WELLS, G. P. (1931). *Science of Life.* London.
WILLIS, J. C. (1922). *Age and Area.* Cambridge
WILSON, E. B. (1925). *The Cell.* (3rd ed.) New York.
—— and Others (1924). *General Cytology.* Chicago.
WOODS, F. A. (1906). *Heredity in Royalty.* New York.

(2) *PAPERS*

AFZELIUS, K. (1924). Senecio. *Act. Hort. Berg.* VIII, p. 123.
AGOL, I. J. (1930). Compound genes. *Anat. Rec.* XLVII.
ARKWRIGHT, J. A. (1930). Bacteria. *System of Bacteriology. Med. Res. Council,* I, p. 115. London.
AVERY, P. (1930). Crepis. *Univ. Calif. Publ. Agric. Sci.* VI, p. 135.
BABCOCK, E. B. and NAVASHIN, M. (1930). Crepis. *Bibl. Genet.* VI, p. 1.
BARNARD, J. E. (1930). Bacteria. *System of Bacteriology. Med. Res. Council.* London.
BATESON, W. (1906). *Rep. Evol. Com. Roy. Soc.* III, p. 31.
—— (1908). *Ibid.* IV, p. 6.
BATESON, W., PUNNETT, R. C. and SAUNDERS, E. R. (1905). Sweet Peas. *Rep. Evol. Com. Roy. Soc.* II, p. 80.
BELLING, J. (1928). Chromomeres. *Univ. Calif. Publ. Bot.* XIV, p. 307.
BLACKBURN, K. B. and HESLOP-HARRISON, J. W. (1924). Salicaceae. *Ann. Bot.* XXXVIII, p. 361.
—— —— (1921). Rosa. *Ann. Bot.* XXXV, p. 159.
BLAKESLEE, A. F. (1927). Datura. *Ann. N. York Acad. Sci.* XXX, p. 1.
—— (1929). Datura. *Journ. Hered.* XX, p. 177.
—— (1930). X-rays (*Datura*). *Anat. Rec.* XLVII, p. 383.
BLAKESLEE, A. F., MORRISON, G. and AVERY, A. G. (1927). Datura. *Journ. Hered.* XVIII.
BRUUN, H. G. (1930). Primula. *Svensk Bot. Tidskr.* XXIV, p. 468.
BURNET, F. M. (1930). Bacteriophage. *System of Bacteriology. Med. Res. Council.,* VII, p. 463. London.
BUXTON, B. H. and NEWTON, W. C. F. (1928). Digitalis. *Journ. Genet.* XIX, p. 269.
CASTLE, W. E. (1905). Rabbits. *Publ. Carnegie Inst. Wash.* 1905. (See Book, 1932.)
CHITTENDEN, R. J. (1928). Godetia. *Journ. Genet.* XIX, p. 285.
CLAUSEN, J. (1926). Viola. *Hereditas,* VIII, p. 1.
—— (1927). Viola. *Ann. Bot.* XLI, p. 677.
—— (1931). Viola. *Hereditas,* XV, p. 219.
CLAUSEN, R. E. (1928). Nicotiana. *Verh. Int. Kongr. Vererb. Berlin,* 1927, p. 547.
—— (1928). Nicotiana (new species). *Univ. Calif. Bibl. Bot.* XI, p. 177.
CLELAND, R. E. (1925). Oenothera. *Amer. Nat.* LIX, p. 475.
—— (1929). Oenothera. *Zeitschr. ind. Abstamm. u. Vererb.* LI, p. 126.
—— (1931). Oenothera. *Proc. Nat. Acad. Sci.* XXVII, p. 437.
CLELAND, R. E. and BLAKESLEE, A. F. (1931). Oenothera. *Cytologia,* II, p. 175.

COLLINS, J. L., HOLLINGSHEAD, L. and AVERY, P. (1929). *Crepis artificialis. Genetics*, XIV, p. 305.

CRANE, M. B. and LAWRENCE, W. J. C. (1929). Cultivated fruits. *Journ. Pom. and Hort. Sci.* VII, p. 276.

—— —— (1930). Apple chromosomes. *Journ. Genet.* XXII, p. 153.

—— —— (1931). Sterility in fruits. *Rep. and Proc. 9th Int. Hort. Congr. London*, 1930.

DANFORTH, C. H. (1929). Skin transplantation. *Journ. Hered.* XX, p. 319.

DARLINGTON, C. D. (1928). *Prunus. Journ. Genet.* XIX, p. 213.

—— (1930). Crossing over. *Proc. Roy. Soc.* B, CVII, p. 50.

—— (1930). *Prunus. Journ. Genet.* XXII, p. 65.

—— (1931). Meiosis. *Biol. Rev.* VI, p. 264.

DARLINGTON, C. B. and MOFFETT, A. A. (1930). *Pyrus. Journ. Genet.* XXII, p. 129.

DAVENPORT, C. B. (1907). Eye colour. *Science*, XXVI, p. 589

—— (1930 A). Sex-linkage in Man. *Genetics*, XV, p. 401.

—— (1930 B). Evolution. *Science Monthly*, XXX, p. 307.

DEMEREC, M. (1926). Maize. *Amer. Nat.* LX, p. 172.

—— (1928). Genes. *Verh. Int. Kongr. Vererb. Berlin*, 1927, p. 183.

DE MOL, W. E. (1926). Polyploids. *La Cellule*, XXXVIII, p. 7.

—— (1928). Tulips. *Nat. Gesellsch. Zürich*, LXXIII, p. 73.

—— (1928). Tulips. *Genetics*, XI, p. 120.

—— (1930). Induced mutations. *Zeitschr. ind. Abstamm. u. Vererb.* LIV, p. 363.

DE VRIES, H. (1929). New mutations in *Oenothera. Zeitschr. ind. Abstamm. u. Vererb.* LII, p. 121.

DIAKONOV, D. M. and LUCE, J. J. (1922). Musical ability. *Bull. Eugen. Bur. Leningrad*, I.

DRINKWATER, H. (1916). Musical ability. *Journ. Genet.* V, p. 229.

EGHIS, S. A. (1930). *Nicotiana. Proc. U.S.S.R. Congr. Genet. Plant and Animal Breeding*, II, p. 571.

GAIRDNER, A. E. (1926). *Campanula. Journ. Genet.* XVI, p. 341.

GATES, R. R. (1928). *Oenothera. Bibl. Genet.* IV, p. 401.

GATES, R. R. and SHEFFIELD, F. M. L. (1929). *Oenothera. Phil. Trans. Roy. Soc.* B, CCXVII, p. 367.

GATES, R. R. and GOODWIN, K. M. (1930). Haploids. *Journ. Genet.* XXIII, p. 123.

GOLDSCHMIDT, R. (1928). The gene. *Quart. Rev. Biol.* III, p. 307.

—— (1931). Intersexes (Moths). *Ibid.* VI, p. 125.

GOODSPEED, T. H. (1929). X-radiation (*Nicotiana*). *Journ. Hered.* XX, p. 243.

—— (1930). Polyploids (X-rays). *Univ. Calif. Publ. Bot.* XI, p. 299.

GOODSPEED, T. H. and CLAUSEN, R. E. (1925). *Nicotiana* (new species). *Genetics*, X, p. 279.

GOODSPEED, T. H. and OLSON, A. R. (1928). X-rays (*Nicotiana*). *Proc. Nat. Acad. Sci.* XIV, p. 66.

GOODSPEED, T. H. and AVERY, P. (1930). X-rays. *Cytologia*, I, p. 308.

HAASE-BESSELL, G. (1916). *Digitalis. Zeitschr. ind. Abstamm. u. Vererb.* XVI, p. 293.

—— (1922). *Digitalis. Ibid.* XXVII, p. 1.

—— (1926). *Digitalis. Ibid.* XLII, p. 1.

HÅKANSSON, A. (1931). *Pisum. Hereditas*, XV, p. 17.

HALDANE, J. B. S. (1929). Polyploids. *Rep. Conf. Polyploidy, John Innes Hort. Inst.* p. 9.

HALDANE, J. B. S. (1930). Polyploid genetics. *Journ. Genet.* XXII, p. 359.

HANSON, F. B. and WINKLEMAN, E. (1929). Radium irradiation. *Journ. Hered.* XX, p. 277.

HEILBORN, O. (1927). *Draba. Hereditas,* IX, p. 59.

—— (1929). Chromosomes and taxonomy. *Proc. Int. Congr. Plt. Sci. New York,* 1926, p. 307.

HESLOP-HARRISON, J. W. and GARRETT, F. C. (1926). Melanism in Moths. *Proc. Roy. Soc.* B, XCIX, p. 241.

HOLLINGSHEAD, L. (1928). Haploid *Crepis. Amer. Nat.* LXII, p. 282.

—— (1928). Chimeras. *Univ. Calif. Publ. Agric. Sci.* II, p. 343.

—— (1930). *Crepis* hybrids. *Univ. Calif. Publ. Agric. Sci.* VI, p. 55.

HOLINGSHEAD, L. and BABCOCK, E. B. (1930). *Crepis. Univ. Calif. Publ. Agric. Sci.* VI, p. 1.

HURST, C. C. (1898). Evolution by hybridisation. *Nature,* LIX, p. 178.

—— (1904). Rabbits, poultry, sweet peas, etc. *Trans. Leic. Lit. and Phil. Soc.* VIII, p. 121.

—— (1905). Poultry. *Rep. Evol. Com. Roy. Soc. London,* II, p. 131.

—— (1907). Eye colour. *Nature,* LXXVI, p. 55.

—— (1908A). Eye colour. *Proc. Roy. Soc.* B, LXXX, p. 85.

—— (1908B). Musical temperament. *Trans. Leic. Lit. and Phil. Soc.* XII, p. 35.

—— (1910). Sweet peas. *Journ. Roy. Hort. Soc.* XXXVI, p. 22.

—— (1913). Dutch rabbits. *Experiments in Genetics,* 1925, p. 456.

—— (1924). Polyploid species. (*Rosa.*) *Rep. Brit. Assoc. Liverpool,* 1923. *Experiments in Genetics,* 1925, p. 534.

—— (1926). Origin of species. *Proc. Linn. Soc. London,* 1925–6, p. 30.

—— (1927A). Species concept. *Rep. Brit. Assoc. Oxford,* 1926, *Science,* March 28, 1927.

—— (1927B). Mechanism of evolution. *Eugenics Rev.* XIX, p. 19.

—— (1928). Polyploidy in *Rosa. Rep. 5th Int. Congr. Genet. Berlin,* 1927, in *Zeitschr. ind. Abstamm. u. Vererb.* Suppl. II, p. 866.

—— (1929A). Polyploidy. *Rep. Conf. Polyploidy, John Innes Hort. Inst.* p. 13.

—— (1929B). Genetics of *Rosa. Rep. Int. Rose Conf.* 1928, in *Rose Annual,* 1929, p. 37.

—— (1930A). Pedigree of Bach. *Eugenics Rev.* XXII, p. 64.

—— (1930B). Species concept. *Gard. Chron.* LXXXVIII, p. 325.

—— (1931A). Speeding up plant breeding. *Emp. Cotton Grow. Rev.* VIII, p. 103.

—— (1931B). Embryo-sacs in *Roseae. Proc. Roy. Soc.* B, CIX, p. 126.

—— (1931C). New species concept. *Rep. Int. Bot. Congr. Cambridge,* 1930.

—— (1932A). Russian experiments. *Emp. Cotton Grow. Rev.* IX, p. 4.

—— (1932B). Genetics of Intellect. *Proc. Roy. Soc.* B, CXII, p. 80.

HUSKINS, C. L. (1928). Speltoid wheats. *Journ. Genet.* XX, p. 103.

—— (1931). Origin of *Spartina Townsendii. Genetica,* XII, p. 531.

HUSKINS, C. L. and SMITH, S. G. (1932). *Sorghum. Journ. Genet.* XXV, p. 241.

KARPETSCHENKO, G. D. (1927). *U.S.S.R. Bull. Appl. Bot. Genet.* XVII, p. 205.

—— (1928). New Genes. *Zeitschr. ind. Abstamm. u. Vererb.* XLVIII, p. 1.

—— (1929). Chromosomes of *Raphanobrassica. Der Züchter,* V (1. Jahrg.).

—— (1929). Three-species hybrids. *Proc. U.S.S.R. Congr. Genet. Plant and Animal Breeding,* II, p. 277.

LAMMERTS, W. (1929). *Nicotiana. Genetics*, XIV, p. 286.

LAWRENCE, W. J. C. (1929). *Dahlia. Journ. Genet.* XXI, p. 125.

—— (1931). *Dahlia. Ibid.* XXIV, p. 257.

—— (1931). Secondary polyploids. *Cytologia*, II, p. 352.

LESLEY, M. M. and FROST, H. B. (1928). *Matthiola. Amer. Nat.* LXII, p. 22.

LESLEY, J. W. (1928). Triploid tomatoes. *Genetics*, XIII, p. 1.

LINDSTROM, E. W. (1929). Tomato. *Journ. Hered.* XX, p. 23.

—— (1930). Maize. *Bull. Torrey Bot. Club*, LVII, p. 221.

LJUNGDAHL, H. (1924). *Papaver. Svensk Bot. Tidsk.* XVIII, p. 279.

LUSH, J. L. (1930). Cattle. *Journ. Hered.* XXI, p. 85.

MARSDEN-JONES, E. M. (1930). × *Geum. Journ. Genet.* XXIII, p. 377.

MARSDEN-JONES, E. M. and TURRILL, W. B. (1930). Tetraploid Saxifrage. *Ibid.* XXIII, p. 83.

McCLINTOCK, B. (1929). Maize. *Journ. Hered.* XX, p. 218.

MJÖEN, J. (1925). Inheritance of musical ability. *Hereditas*, VII.

—— (1926). Inheritance of musical ability. *Eugenics Rev.* 1926.

MOFFETT, A. A. (1931). *Pomoideae. Proc. Roy. Soc.* B, CVIII, p. 423.

MORGAN, T. H. (1927). Biology and physics. *Science*, LXV, p. 213.

MORGAN, T. H., BRIDGES, C. B. and SCHULTZ, J. (1930). *Carnegie Inst. Year Book*, XXIX, p. 352.

MORGAN, T. H., BRIDGES, C. B. and STURTEVANT, A. H. (1925). *Drosophila. Bibl. Genet.* II, pp. 1–262.

MULLER, H. J. (1927). X-radiation. *Science*, LXVI, p. 84.

—— (1929A). Genes. *Proc. Int. Congr. Plt. Sci.* I, p. 897.

—— (1929B). Translocation. *Amer. Nat.* LXIII, p. 481.

—— (1930A). Variations induced by X-rays. *Journ. Genet.* XXII, p. 299.

—— (1930B). X-radiation. *Amer. Nat.* LXIV, p. 220.

MULLER, H. J. and ALTENBURG, E. (1930). *Genetics*, XV, p. 283.

MULLER, H. J. and NOTT-SMITH, L. M. (1930). Radio-activity and mutations. *Proc. Nat. Acad. Sci.* XVI, p. 277.

MÜNTZING, A. (1928). Pseudogamie. *Hereditas*, XI, p. 267.

—— (1930). *Galeopsis. Ibid.* XIII, p. 185.

—— (1932). *Galeopsis. Ibid.* XVI, p. 105.

NEWTON, W. C. F. (1926). *Tulipa. Journ. Linn. Soc.* XLVII, p. 339.

NEWTON, W. C. F. and PELLEW, C. (1929). *Primula kewensis. Journ. Genet.* XX, p. 405.

ONO, T. and SHIMOTOMAI, N. (1928). *Rumex. Bot. Mag. Tokyo*, XLII, p. 266.

OSTENFELD, C. H. (1921). *Hieracium. Journ. Genet.* II, p. 117.

PAINTER, T. S. and MULLER, H. J. (1929). Transmutations by X-rays. *Journ. Hered.* XX, p. 287.

PATTERSON, J. T. (1929). X-ray mutations. *Journ. Hered.* XX, p. 261.

PATTERSON, J. T. and MULLER, H. J. (1930). Progressive mutations by X-rays. *Genetics*, XV, p. 495.

PELLEW, C. and SANSOME, E. R. (1931). *Pisum. Journ. Genet.* XXV, p. 25.

PHILIPTSCHENKO, J. (1927). "Marquis" wheat, and musical ability in man. *Bull. Bur. Genet. and Plt. Br. Leningrad*, V.

PICKARD, J. N. (1929). Rabbits. *Journ. Hered.* XX, p. 483.

PUNNETT, R. C. and PEASE, M. S. (1925). Dutch rabbits. *Journ. Genet.* XV, p. 375.

RICHARDSON, E. (1929). *Pisum. Nature*, CXXIV, p. 578.

ROSENBERG, O. (1906). *Hieracium. Ber. deutsch. Bot. Gesell.* XXIV, p. 157.

—— (1930). Parthenogenesis. *Handb. d. Vererbungswissensch.* II.

SACHAROV, V. (1924). Music pedigrees. *Russ. Eugen. Journ.* II, p. 2.

SALAMAN, R. N. (1932). Virus complexes. *Proc. Roy. Soc.* B, CX, p. 186.

SANSOME, F. W. (1929). Tomato. *Rep. Conf. Polyploidy, John Innes Hort. Inst.* p. 45

SAX, K. (1930A). *Rhododendron. Amer. Journ. Bot.* XVII, p. 247.

—— (1930B). Crossing over. *Journ. Arnold Arbor.* XI, p. 193.

SCHIEMANN, E. (1930). *Fragaria. Ber. deutsch. Bot. Gesell.* XLVIII, p. 211.

SEREBROVSKY, A. S. and DUBININ, N. P. (1930). *Journ. Genet.* XXI, p. 259.

SHIMOTOMAI, M. (1931). Chrysanthemum, I. *Journ. Sci. Hiro. Univ.* I, p. 37.

—— (1932). Chrysanthemum, II. *Ibid.* I, p. 117.

SIMONET, M. (1928). *Iris. Compt. rend. Soc. Biol. Paris*, XCIX, pp. 1314, 1928.

SKOVSTED, A. (1929). *Aesculus. Hereditas*, XII, p. 64.

SMITH, HENDERSON (1927). Plant viruses. *Proc. Roy. Soc. Med.* XX, p. 11.

—— (1930). Intracellular bodies. *Nature*, Feb. 8, 1930.

SMITH, KENNETH (1931). Potato *Y* virus. *Proc. Roy. Soc.* B, CIX, p. 251.

STADLER, L. J. (1930). X-radiation. *Journ. Hered.* XXI, p. 3.

STOUGHTON, R. H. (1929). Bacteria. *Proc. Roy. Soc.* B, CV, p. 469.

STROGAYA, E. (1926). Musical ability. *Russ. Eugen. Journ.* IV, p. 2.

STURTEVANT, A. H., BRIDGES, C. B., MORGAN, T. H. and Others (1929). *Drosophila* species. *Carnegie Inst. Wash. Publ.* 399.

TACKHOLM, G. (1922). *Rosa. Act. Hort. Berg.* VII, p. 97.

TAYLOR, W. R. (1925). *Gasteria. Amer. Journ. Bot.* XII, p. 219.

—— (1931). *Gasteria. Rep. 5th Int. Congr. Bot. Cambridge,* 1930.

TIMOFEEFF-RESSOVSKY, H. A. (1930). *Drosophila funebris. Journ. Hered.* XXI, p. 167.

—— (1931). Reverse mutations. *Ibid.* XXII, p. 67.

TISCHLER, G. (1927). Chromosomes. *Tabl. Biol. Period.* IV, p. 1.

—— (1931). *Ibid.* VII, p. 109.

TSCHERMAK, E. and BLEIER, H. (1926). *Aegilotricum. Ber. deutsch. Bot. Gesell.* XLIV, p. 110.

—— (1930A). *Triticum. Rep. Int. Bot. Congr. Cambridge,* Abstracts, p. 152.

—— (1930B). *T. turgido-villosum. Ber. deutsch. Bot. Gesell.* XLVIII, p. 400.

VAVILOV, N. S. (1922). Origin of species. *Journ. Genet.* XII, p. 47.

—— (1930). Origin of fruit trees. *Rep. and Proc. 9th Int. Hort. Congr. London,* p. 271.

—— (1931). Linnean species. *U.S.S.R. Bull. Appl. Bot. Genet.* XXVI, p. 109.

—— (1931). Origin of cultivated plants. *Ibid.* XXVI, p. 135.

VILMORIN, R. DE and SIMONET, M. (1928). *Solanum. Verh. Int. Kongr. Vererb. Berlin,* 1927, p. 95.

WATKINS, A. E. (1930). Wheat species. *Journ. Genet.* XXIII, p. 173.

—— (1932). Hybrid sterility. *Journ. Genet.* XXV, p. 125.

WHITING, P. W. (1929). X-rayed Wasps. *Journ. Hered.* XX, p. 269.

WINGE, Ö. (1930A). *Lebistes* (Fish). *Journ. Genet.* XXIII, p. 69.

—— (1930B). Chromosomes of mouse tumours. *Zeitschr. f. Zellforsch. u. mikro. Anat.* X, p. 683.

WOODWORTH, R. H. (1929). Chromosomes of *Betula. Bot. Gaz.* LXXXVII, p. 331

(References to other important papers will be found in the Books listed above.)

INDEX

Printed in the United States
By Bookmasters